A Village Blacksmith

The Story of the Family Business of G.S. Whiteley and Co. of Ogden Lane, Rastrick, Brighouse, Yorkshire

by
Clifford Riley

Bloomington, IN Milton Keynes, UK

AuthorHouse™
1663 Liberty Drive, Suite 200
Bloomington, IN 47403
www.authorhouse.com
Phone: 1-800-839-8640

AuthorHouse™ UK Ltd.
500 Avebury Boulevard
Central Milton Keynes, MK9 2BE
www.authorhouse.co.uk
Phone: 08001974150

This book is a work of non-fiction. Unless otherwise noted, the author and the publisher make no explicit guarantees as to the accuracy of the information contained in this book and in some cases, names of people and places have been altered to protect their privacy.

© 2007 Clifford Riley. All rights reserved.

No part of this book may be reproduced, stored in a retrieval system, or transmitted by any means without the written permission of the author.

First published by AuthorHouse 5/31/2007

ISBN: 978-1-4259-9972-8 (sc)

Printed in the United States of America
Bloomington, Indiana

This book is printed on acid-free paper.

*My mother, Ruth; my father, John;
me as a child*

Acknowledgements

I should like to thank Judith for collating my random jottings, for the photographs and for her encouragement; Kathryn for her hours of typing; Wendy for her photographs and her contribution to the penultimate chapter; and to Julie for her computer work, patience and help throughout in the writing of this book.

Clifford Riley – Summer 2006

About the Author

Clifford Riley is a man who knows his trade and takes great pride in it. His career as a Blacksmith in the family firm of G.S. Whiteley's, Rastrick, Brighouse, which spans much of the 20th century, is also a story of the events and changes that affected the inhabitants of this little West Yorkshire town. After 86 years of living within two miles of the smithy, he has become a well-known and much respected local character, not the least because of his integrity, gentleness and commanding presence.

This book contains a wealth of technical information and drawings about the work of a smith, written with a clarity and precision of language which can be accessed by all. It is interspersed with amusing and witty anecdotes, introducing us to a world rapidly disappearing under the onslaught of 'progress'.

For those who enjoy reading about social history, this book is a must. It combines Northern landscapes, Northern grit and Northern humour with the natural storyteller's eye for detail and the craftsman's skill. In conclusion I quote Clifford, who writes at the end of this book, "When my time comes it will be the end of an era. I only hope people will remember me as a fellow who did a good job".

Table of Contents

Acknowledgements vii

About the Author ix

Chapter 1 – My Early Life in Rastrick 1
    Ogden Lane 15

Chapter 2 – Mrs Parrat 17
    The White House 19

Chapter 3 – A Blacksmith's Apprentice 20
    Pitching Tool or Pitch Nicker 28

Chapter 4 – Rastrick Church Bell and Communion Rail 29
    Stone Nicker 32

Chapter 5 – Hooping 33
    Brick Nicker 38

Chapter 6 – The Factory Inspector 39
    Holeing Pick and Splitting Wedges 41

Chapter 7 – Rastrick Illuminations 42
    Plugs and Feathers 45

Chapter 8 – Local Transport 46
    Lifting Lewis 49

Chapter 9 – The shearing machine 50
    Dummy Dog, Saw Frame Dog and Cotter 51

Chapter 10 – Siege Bands 53
    Wire Rope End Socket 57

Chapter 11 – Pointing Irons 58
    Rock Dog and Nip Dog 61

Chapter 12 – The weather vane..62
    Lifting Wedges...65

Chapter 13 – Sladdins...66
    Claw Tool Holder 68

Chapter 14 – The Riley Elevator Truck 69

Chapter 15 – Wrought Ironwork 72

Chapter 16 – Odd Jobs...77

Chapter 17 – War work...85

Chapter 18 – Home Guard Duty 90

Chapter 19 – A Woolworth's Hacksaw Blade..98

Chapter 20 – Nora.. 109

Chapter 21 – From Bikes to Vans.. 112

Chapter 22 – My Father... 120

Chapter 23 – Ministry of Supply 127

Chapter 24 – The Riley Link... 131

Chapter 25 – ICI work 136

Chapter 26 – The Flood 144

Chapter 27 – The French Connection..147

Chapter 28 – John 150
    Two Pictures of the Smithy, about 70 years apart 155

Chapter 29 – Nora's Death 156

Chapter 30 – Uncle George 158

Chapter 31 – The 'Bending' machine 164

Chapter 32 – Emmie 167

Chapter 33 – Praise Indeed 169

Chapter 34 – Retirement 171

Chapter 35 – Smithy's End. 175

Chapter 36 – Finale 182

Glossary of Tools 187

Chapter 1

My Early Life in Rastrick

"Under the spreading chestnut tree the village smithy stands. The smith a mighty man is he with large and sinewy hands."[1]

Those of you who know me will be aware that I spent my working life in the smithy at Ogden Lane, Rastrick, first started by my father's uncle George in 1860. A chestnut tree actually grows in the golf links just over the wall, but as most of the conkers fall on our side, we can lay some claim to it and apply the quotation from this well-known poem to our firm.

The Smithy c.1990, showing the trees over the wall

---
[1] The Village Blacksmith (Henry Wadsworth Longfellow)

Clifford Riley

I think I must have always been intended to be a blacksmith. I have a photograph showing me as a toddler at home, in a little white frock, standing on a chair, and anyone who sees it remarks about my large, chubby hands even at that age. I still have the large hands and have been known to lift a 3cwt [2] anvil onto its block.

Clifford aged three

My parents were both musicians, but when I was born, my father, a cellist, saw me as a blacksmith and my mother, a violin teacher, saw me

---

[2] Hundredweight – imperial measure

as a musician. Unfortunately my mother died when I was three, so I was destined to become a blacksmith.

My mother

My father always encouraged me to potter about in the smithy – there were no health and safety issues in those days – and I was fascinated by the fire. Often at dinner times I would go and blow the fire up with the bellows and get it started for the men who came back to start work again at one o'clock. After a time I became more adventurous and started putting bits of scrap metal into the fire and getting them red hot.

During my years at Carr Green School nearly every boy had a bully bowl [3] and handle and you used to hang these on your hook in the cloakroom, along with your cap and coat. Some lads did not have a proper hook, but

---

[3] large hoop to roll along the ground

propelled their hoops with a stick, so I got many requests for the requisite hook, which they hoped that I would be able to produce at no charge.

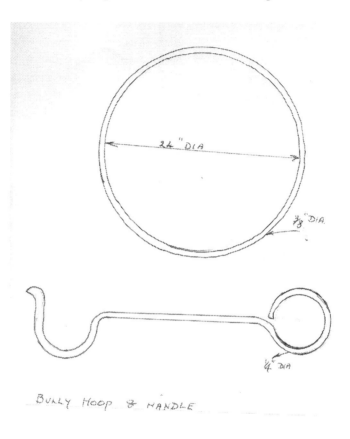

Drawing of a bully bowl hoop and hook

I had watched the apprentices make these hooks, so one dinnertime I decided to have a go myself. I made my first hook at the age of ten. As I went along I improved, but one day, as I was putting the finishing touches to one of my best hooks, it slipped off the anvil pike and gave my thigh a glancing blow. I was left with a rather painful brand! When my father saw it he complimented me on the shape of the handle!

After school, if I ever went into the smithy, there was always a job to do. One was to fetch buckets of coke for the men to put on the hearths. If I fetched the coke it meant that they had more time on the anvil and I was taught from an early age that 'time was money'.

A fire at the forge

Saturday afternoons, for father and me, were spent on maintenance. We had a Crossley oil engine to run the line shafting and every week certain plates were removed so that carbon deposits could be scraped away and the ports made clean. All the oil bottles on the line shaft had to be filled with oil too. Every bearing had a little bottle of oil stuck in the top with a spindle running through the cork and onto the running shaft. This system kept the bearings lubricated. The power saw, grinding machines and power hammers, which all had holes for oil, had to have the requisite squirt applied also.

Inside the shop c.1990

Living as we did on Stackgarth gave us a grandstand view of one Rastrick event of the year. This was the motorcycle hill climb up the field known as the Brow. Bikes were not as powerful as today and it was an achievement to get to the top. There were marshals with ropes with hooks on the end to stop anyone careering back to the bottom if they stalled the engine. Over the years, as engines became more powerful, they dug a ditch at the bottom to stop anyone getting a racing start.

Often at the weekend we heard the sound of gunfire as the Territorials practised shooting on the range between Toothill Bank and Round Hill. There were platforms at 100 and 200 yards range, the last named being on the left of Toothill bank near where the White House used to be. From New Dick (between the top of Toothill Bank and Clough Lane) you got a grandstand view and could see all the shots being signalled in the butts. I never thought that in later years I should be shooting there myself with the Home Guard.

Round Hill and Rastrick Ground, c.1920

I started school in 1926. Going to school was enjoyable, as there always seemed to be a gang of us going in clogs and returning at the same time. Wearing clogs was wonderful. They were hardwearing and warm on the feet; also they made a very loud noise. Cobbled roads were easier to manage with clogs because your feet were protected. The only problem was on very snowy days when the snow stuck to the metal on the bottom of the clog and you were soon walking on what seemed like stilts and had to kick off the build-up of snow.

As I look at that journey now, how things have changed! The Junction Pub is still there, but the grammar school has gone; the Thornhill Arms is now a Nursing home; Mrs Parrat's, the Co-op, the fish shop and the chemists are all gone; the chemists used to be on the left-hand side of Church Street and J. A. Binns was the Dentist's opposite. Older lads used to say that when he pulled a tooth you finished standing in the chair. We were terrified of getting toothache. The paper shop is still there.

The Junction Pub from the Smithy and looking up Ogden lane from the Smithy

At the end of Church Street, opposite the church and at the junction of Jumble Dyke (or 'top of the town' as it was known) was Martins Bank, later to become a paint shop. The branch was only open on Tuesday and Friday for the convenience of the industry in Rastrick. My father used to go every Friday for the wages and he always changed his bowler hat first. I used to think that going to the bank for money was a good idea, until one day, when I decided to go for some myself and my father explained to me that you had to put it in first. I could never see the sense of that!

Church Street, Rastrick, 'Top of town', c.1900, looking towards the Grammar School

Opposite the bank, adjacent to the church, where the church car park is now, was the Lower George Public House and a little bit higher up came Clays Mill where some of the finest cloth in the land was woven. This was later to become Sladdins who made shoulder pads.

Back across the road again stood the Crowtrees laundry, later to move to Brighouse, but now gone completely. At the bottom of the grass verge, where the road divides, there was an old cannon, a relic from the 1914 war. Thank goodness the library has survived; not so Crowtrees Chapel, which stood opposite Carr Green School and now has houses on the site. The old School is now a Nursing home.

Crowtrees Lane, c.1930, Showing World War I cannon and Library to the right

At the corner of Carr Green lane was Carr Green Well. It was really a horse trough where water flowed in from a pipe at one end and out at the other end. It was about 6ft long, 3ft wide and 4ft high. The walls were about 4" thick and this was hewn from a solid block of local stone. Just consider the work involved in chiselling out these troughs.

If you lived above the school you were a top-ender; we were bottom of towners, bottom of town being between Bowling Alley and Toothill Bank. Church Street was known as the top of town. Above the school there was Helms Mill, also noted for fine cloth; Ramsden's toffee and sweet works, and Smith's Mill. The Smith's Orphanage at Boothroyd was financed by the mill.

Delph Hill, Rastrick, c.1920; the woodworking shop of Hiram Hill was on the left where the children are

Still going strong is New Road Sunday School. New Road Sunday School was unique in so far as it was a Sunday school only and was non – denominational; it was neither, C of E, Methodist, Catholic, nor any other sect. The choir was always called the singing class. The institute, as it was known, was an integral part of the Sunday school building, but had a separate entrance so it could be used without opening up the schoolroom, and it housed a full-sized billiard table. They always had a good cricket team with their own field and pavilion and played in the Huddersfield League. Every year they put on a musical operetta with an orchestra made up of local musicians and these occasions were well supported. My father attended New Road and was a Sunday School Superintendent, playing his cello in the orchestra, so I was always privileged to have a front row seat near him. They had a strong Band of Hope, which was held twice a month, where we were taught to, "Beware the Demon Drink". We sang:

> My drink is water bright,
> Water bright, water bright,
> From the crystal spring.

And,

> Dare to be a Daniel,
> Dare to stand alone,
> Dare to pass a public house
> And take your money home.

I later learned that my Great Grandfather was a big drinker and left my Grandfather with a lot of debts to pay off, so it was not surprising that our family was strictly teetotal. My cousin Norman would not eat a wine gum and was later to take high office in the Band of Hope movement. Imagine the furore, then, when his brother Harold married the daughter of the landlord of the Lower George Pub, which was situated where the church car park is now. She was a lovely lass, called Marjorie Huggett, who unfortunately died a few years later, leaving a little daughter, Elaine, who was to grow up and serve for many years as an officer in the Salvation Army.

New Road Sunday School was the hub of the top end of Rastrick and it was fitting that the premises were also used for a British restaurant during the war years. They served hot meals for the many workers from the textile mills in Rastrick.

Sometimes we left school to return through the fields, down Carr Green Lane and past the cemetery. If we had plenty of time we might go along the path at the top side of the cemetery where at the corner of the wall there was a bubbling well where the water used to ebb and flow. We used to like to listen to it. Then we walked through the recreation ground and back down Quarry Road, to the bottom of Toothill Bank. From the 'Rec' a stream runs down the side of the field and it never dries up. At the bottom of Toothill bank it goes underground; under The Junction Pub and under the Golf Links; where it then flows past the Dye Works into Walshaw Drake's Dam (now demolished); disappearing again to join the river, just behind the shops by the bridge, at the bottom of Bridge End.

Castlefields Drive, c.1920

There must be a lot of water in Rastrick, because there were springs running into troughs at Clough Lane, New Dick and by the cricket field at Badger Hill.

Coming back to Quarry Road; when the houses were demolished some years ago (where the bowling club car park is now), a water wheel was found underneath the floor of the kitchen of the end house, which must have been fed by the stream. It was going to cost too much money to remove it, so it was buried where it was. Mr and Mrs Walton and their large family lived there. Many, many years ago, we are told, there was a tannery on the site.

I eventually went to Rastrick Grammar School and endured the initiation ceremony. This consisted of:
- Having your head put into the mouth of a cannon whilst it was hammered with sticks and stones;
- Being 'possed'[4] on a tree stump;
- Being dropped over a wall into the rubbish enclosure;
- Made to run the gauntlet of the rest of the school all lined up with sticks;
- Being put down the glory hole (school cellar) as it was called, and your head put under the cold tap.

Somehow everyone survived!

Rastrick Grammar School, on the left, and The Thornhill Arms, c.1930. Now the Cottage Nursing Home

---

[4] Moved up and down like a posser in washing tubs. 4 boys, each holding a leg or an arm, lifted you up and then lowered you, so that your backside landed on the stump.

Ogden Lane

This is an old picture of Ogden Lane and the Smithy was just at the bottom on the left. Halfway down is the entrance to Stackgarth, where I was brought up and lived for 29 years. The Junction Pub is at the bottom right. Top left is Mrs Parrat's sweet shop with the blind over the window.

Between the shop and Stackgarth was a municipal property and the noticeboard outside the top door is quite visible. On the end of Stackgarth can be seen the old gatepost with the rounded top and just above it a small half door. I never saw this open, but years and years ago this part was known as the 'Old Towser', where the local drunks and miscreants were locked up for the night.

Beyond the houses at the bottom can be seen the hill known as 'The Brow', where the motor cycle trials were run. At the bottom of the hill outside the Junction Pub can be seen the gas lamp in the middle of the road. Two horse drawn carts can also be seen. Because the hill was steep and setted it was very difficult for the horse. The carts carried a block of wood on two chains and these were hooked up at the bottom of the hill and dragged behind the back wheels. If the horse tired or slipped the cart was stopped from running backwards by the wood blocks. The horse could also stop for a rest at the end of Stackgarth.

Chapter 2

Mrs Parrat

Everyone knew Mrs Parrat, who had the little shop in Ogden Lane. Along with her daughter she sold sweets of all descriptions and baked parkin and jam pasties. The jam pasties were white and thin. The school boys said they had been through the mangle and the jam was a thin red line. The parkin, however, was good and for many, many years my supper was a bottle of milk and a square of Mrs Parrat's parkin.

She was very fond of cats and on sunny days there was always one in the window curled up among the boxes of sweets.

Mrs Parrat had a little back room which had a little window that opened onto the Grammar School playground. As the boys were not allowed out of the playground during school hours it was convenient to reach over the wall and purchase sweets which Mrs P. would pass through this window. She was very careful to get the money first before she handed anything over. Her daughter never married, but when her mother died she assumed the title Mrs Parrat, which had been the cry at the back window for many generations of boys at the Grammar School.

Mrs. Parrat's back door, taken from the Grammar School yard, formerly the Headmaster's House, c.1930

The White House

This old photograph is The White House, a well known landmark at the top of Toothill Bank. The road can just be seen on the left. Just above the road, left of the roof, can be seen the stone quarry with the crane. This is now Chapel Croft Estate and Doctors' Clinic. Just after the war it was the site of 'prefabs' (prefabricated houses, which came complete on the back of a wagon). That area was known as the 'Rubob Oil' (Rhubarb Hole) and from here stone was sent to London for the Royal mews and also used in the construction of Blackpool Tower Building (that is according to local legend).

Chapter 3

A Blacksmith's Apprentice.

I started work on the Wednesday following Easter Tuesday in 1937. One of my first jobs was to take some Hod fittings to John Francis Brown's, Ironmongers, in Brighouse. This firm had a reputation for being rude to customers, but usually had what you needed. The shop was never referred to as Brown's; for some reason everyone gave it its full title.

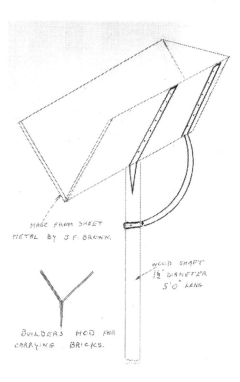

Hod Fittings

Mr Brown was a heavily-built man with a nicotine-stained moustache. Hod fittings were bulky and hard to pack, so to save space I was to carry these in an old sugar sack; they bulged out all over the place. The sack was very uncomfortable on my back and by the time I had carried my load the mile into town, I was ready for dropping it as I entered the shop.

"Mind that glass in the door", bellowed a voice.

All the customers turned to see who was the victim of the proprietor's wrath. One man winked at me, as much as to say "this is quite usual". I handed in my delivery note. Pause.

"Bring 'em over here and let me see 'em", boomed the voice.

I did as ordered.

"Aye, they look alright; put 'em in yon corner."

This I did and prepared to leave. Meanwhile, he had been looking at me with some interest, over the top of his glasses.

"Are you John Riley's lad?" he asked.

"Aye", I said.

"When did you start working for your father?"

"Today", said I.

He paused and then, still looking over his glasses, said,

> "This is the worst day's work tha'll ever do, workin' for thi' father. Who's next?"

A customer stepped forward and normality was restored.

When you are new to the workplace, you soon fall victim to various leg pulls. The first one was, "Go and see Joe for a long stand". Off I went to see Joe.

> "Trevor has sent me for a long stand", I said.

> "Right, wait here", he said and disappeared to another part of the shop.

After a while I was beginning to get a bit suspicious, when he returned, laughing.

> "I think tha's stood long enough now!" he said, and everyone joined in the mirth.

On a visit to Drury's soap works, with some foundation bolts we had forged for fixing a new machine, I fell into the trap again.

> "You are new aren't you?"

> "Aye."

"Welcome to Drury's", he said.

He held out his hand to shake hands. What a change from the reception I got at John Francis Brown's, thought I, and extended my hand. As we clasped hands there was a squelching sound as a lump of soft soap was squeezed into my palm. What a task I had wheeling the barrow back to work with one slippery hand.

I soon found out I was at everyone's beck and call. Every morning each hearth had to be filled with coke; with 6 hearths this took a long time. Mid morning was the time to collect the milk money and go across the road to Bowling Alley and to Pearson Marsdens Dairy for the requisite bottles of milk. There were no tea breaks in those days and a swig of milk, when required, was the only refreshment available.

Although there were three power hammers there were still many jobs that required a striker and one had to quickly learn how to swing a sledge hammer and be available when called.

The worst job for a blacksmith's apprentice apart from striking with the sledge hammer was the delivery and collection of chisels to the stone stand at the end of Birds Royd. Blocks of stone and rough hewn flagstones were brought to the railway sidings from local quarries, where they were worked on and then loaded straight onto rail trucks for dispatch. Our method for transporting the chisels to the stone stand was by means of a wheelbarrow. Unfortunately the road was paved with granite setts, or cobbles, all the way from Ogden Lane to Birds Royd and River Street, not to mention the tram lines, with a loop line for passing, near Jackson's joiners' shop, all of which made the journey difficult. I started with an old wooden barrow, which soon fell apart. As you can imagine the wheel took a battering from the setted road and it was difficult to keep the metal tyre on the wooden wheel.

One day, as I struggled back with the tyre hanging off, Dick Kenworthy, who had a wheelwright's shop down Rastrick Common, came to his gate and couldn't contain his laughter.

> "Tell your father I'll make him a new turl[5] for 12/6."

Dick was a real character and an expert craftsman. The wheels and carts he made were the best to be had and the stringing[6] on the carts was a joy to see. My father was not impressed by the offer of a wooden wheel and he decided that we would make our own metal barrow. This we did complete with a metal wheel with six spokes. What a disaster it turned out to be! The tins of chisels in the metal barrow made a terrible din as I proceeded down Rastrick common; of course Dick was at his gate to enjoy my embarrassment. Worse was to come. The pounding of the setts flattened the rim between the spokes and by the time I got back the wheel was hexagonal. Dick soon saw what had happened.

> "Tell thy father I'll make him a new turl for 15/-," he said as he choked with laughter.

Father, not to be outdone, made a new wheel. This time he welded a disc on each side of the tyre so that it could not flatten. This did the trick, but made the noise worse!

---

[5] A small wooden wheel with a metal tyre
[6] Elaborate paintwork like a sign writer would do

On the journey down Rastrick Common, on the right, was the site of the old Brighouse Brickworks. It is now the site of Millers Oils. This was where the annual fair came, with coconut shies, swing boats, roundabouts and all the usual fun of the fair. The swing boats were like small canal barges which hung from a round shaft high up in the air. They were worked by a sort of brake mechanism which grabbed the rotating shaft and then let go; gradually the boats swung higher and higher, until the floor was almost vertical. Sometimes the lads would drop a coconut on purpose and it would rapidly roll or fall from one end to the other causing mayhem.

On the opposite side of the road there was a stone saw yard where random blocks of stone were sawn up into slabs of stone approximately 2" to 3" thick for dressing into flagstones for pavements.

Further along my route on the left-hand side of Gooder Lane was William Bradley's Engineers. They not only made cast-iron castings and did machining of metal parts, but they made the cranes for the local quarries and all the spares that were needed to keep them in good order.

My journeys to the stone stand in Birds Royd were bad enough. But it was another half mile down River Street to Drury's soap works; and this I dreaded. At the bottom of Mission Street (joining Birds Royd and River Street) one could always take in the smell of leather from Fairburn's Leather works. They used to supply us with leather belting and laces for joining them together.

Eventually I was to get a new barrow with a solid rubber tyre. That was real progress!

Although many boys, and adults too for that matter, watched us at work through the open door, few people outside the trade had any idea what we actually made. Our work was very closely related to the stone trade, as there were many quarries in the area.

Most stone is quarried, in the true sense of the word, where you finish up with a great, gaping hole in the ground; but, apart from Sam Gledhill's (Lower Edge Road) and Paul Normington's (Badger Hill) the Rastrick stone was mined in a similar way to coal. The old shafts are still to be seen adjacent to Ogden Lane, behind 'The Junction' and in the fields near Carr Green new school. These can be clearly seen from the top of Toothill bank. As I mentioned earlier, there was another quarry also, near Chapel Croft, and known as the 'Rhubarb Hole'. Stone from this quarry and stone from other local quarries was supplied for the foundations of Blackpool Tower round about 1890. Many flagstones from Rastrick were supplied to Buckingham Palace and also shipped to Hamburg. All the flags for the Palace had to be 36" x 24" x4".

Most working men enjoyed an opportunity for a clever joke at the expense of new and naïve apprentices; and Blacksmiths were no exception. Apprentices from the quarries came in every day to bring the blunt chisels to be sharpened and take back the newly sharpened ones. They always used this as an opportunity for a smoke and, never having any matches, they would ask the smith for a light. He would light his own cigarette by holding the red hot chisel, in a pair of tongs, half an inch away from the end and lighting the cigarette with it: but he never offered the young lads a light in this manner. Instead, the smith would take another pair of tongs, select a dull-red piece of coke from the edge of the fire and give him that. Because it was a dull red the lad had to take in a very deep breath to get a light in this manner, thus filling his lungs with sulphur from the coke. While he was coughing, the smith would touch the buttons on his jacket with the red-hot chisel, burning the threads and causing the button to fall off. Before the week was out, all his buttons were missing! That is one reason why all the lads had their jackets fastened with a piece of string around the waist.

Stone in Rastrick was mined and was brought to the surface through a round shaft about 12 feet in diameter. The state of the art signalling system to the crane driver was simply a heavy hammer on a lever, with a rope dropping down the shaft. A metal plate was placed under the hammer and one, two or three blows told him all he needed to know.

A new apprentice at the quarry was learning how to 'pitch' the edge of a stone, using scrap pieces of stone that were leftovers. He was of a nervous disposition and every time the hammer clanged on the metal plate he jumped and, on more than one occasion, hit his finger. Not realising the significance of the hammer blows, he found an old sack and put it under the hammer. He was later chased round the quarry top!

Many items in the trade have confusing names and the term 'Dog', for instance, can refer to all kinds of tools. We have a 'saw dog', 'dummy dog', 'nip dog', 'rock dog', 'roof dog' and 'dog grate'. Feathers are not light and could weigh several pounds. Lewis is not a man's name, but a lifting device. Masons Nickers are not to wear, but to split stone.

Pitching Tool or Pitch Nicker

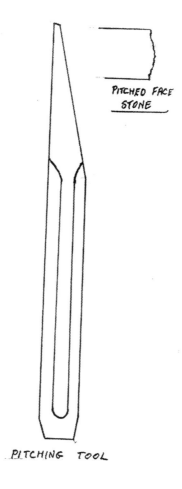

These were made from ⅞" octagon and were 9" long finished. The blade was 2 ½" wide. This meant that the bar had to be jumped up at the end first so that there was enough material to spread. The point was sharpened at an angle, so that instead of splitting the stone, it scalloped the edge and left a rough face. This was known as pitch face building stone.

Chapter 4

Rastrick Church Bell and Communion Rail

St Mathew's Church, Rastrick, c.1930

I had only been working a short while when the vicar came to enquire if we could have a look at the church bell. The beam which held it was cracking. Along with my father I went up into the belfry and took the necessary measurements to make some clamps to hold it together.

A week later, complete with clamps, tool bag and various items of tackle, we made our way up the narrow stairs into the belfry. There was very little working space and father impressed it upon me that I should be very careful and not strike the bell with my hammer or the metal brackets.

"We don't want all Rastrick to hear if we're clumsy", he said.

Suddenly the bell chimed and set our ears ringing.

"I told you to be……"

That was as far as he got as the clock chimed 10 'o clock!

It was then that my father told me the story of how the church bell had saved a man's life.

It was in 1812 when the Luddites were trying to sabotage the new machinery coming into the textile mills. They were of the opinion that the machines would take over their jobs. On the 11th of April that year a raid was organized to attack Cartwright's Mill near Mirfield. The details had been worked out in the local pubs, but too many people heard about it. A Rastrick man named Gaynor was one of the gang and he made his way along Wakefield Road toward the Dumb Steeple at the junction with Leeds Road.

The gang assembled, but they had only gone a short way when they walked into a trap. The police and military were waiting for them. Gaynor turned round, escaped and ran over Bradley Plains to Brighouse and, without stopping, arrived in Rastrick at midnight. As he passed the church gate the sexton was there. It happened that the church clock had been repaired that day and the sexton had just been to check that everything was alright. Gaynor, pausing for breath to pretend he had not been running, stopped to have a word and, as they chatted, the clock struck midnight, but on this occasion it struck 13 times.

Later that night the police arrived to arrest Gaynor, but he protested his innocence and told of his conversation with the sexton at midnight, swearing that he could not have been in Mirfield and back in Rastrick in such a short time. The sexton corroborated his story and, as it seemed impossible for anyone to have run so far so quickly, he was found not guilty. The rest of the gang were hanged.

Another job at the church involved the wrought iron railings in front of the altar. Originally they were in a straight line, but when the church was renovated it was decided to enlarge the podium and make the front curved. The original railings were in good condition, so it was suggested that I should incorporate them in a new curve, matching the new with the old. This was quite a task, but I am very pleased with the result and I defy any member of the congregation to tell which ones are new and which are old.

Interior of St Mathew's Church, c.1930

Stone Nicker

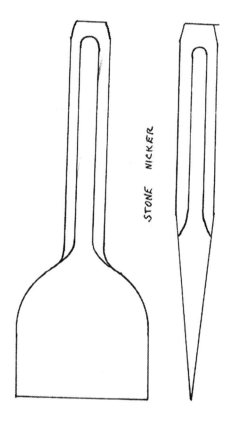

These nickers were made for splitting smaller stones. They were made in different sizes. They were usually 8" – 9" long and 2", 2 ½" and 3" wide. The 2" ones were made from ¾" octagon, the 2 ½" ones from 7/8" octagon and the 3" ones from 15/16" octagon. They were well drawn out and needed to be kept sharp.

Chapter 5

Hooping

I started work in the smithy in 1937, aged 16 and earning the standard wage of 10/-[7] per week for 48 hours work. One of my first recollections of work was the puzzle of why blacksmiths wore white trousers. At least, they were white when they were new. They were made from a sort of cotton twill called fustian. They had no creases and seemed to hang perfectly round in shape. They were very strong-wearing but soon assumed a dark shade. I never had any myself, but the old blacksmiths used to say that on cold winter days you could stand them up in the corner of the bedroom, they were so stiff. True or false? I don't really know.

I was reminded of all this when I saw the old photographs of the 'hooping', which entailed making a metal hoop to fit the cart wheel that the wheelwright had made, or simply replacing a worn-out tyre. We never shod horses, but did make ironwork for carts and specialised in hooping.

---

[7] Ten shillings old money!

Here my father and Joe Vickers were obviously wearing fustian

When the cartwheel came in from the wheelwright it had been put together in sections with close fitting joints. There were no nails or screws to hold it together, just the tightness of the steel tyre (hoop). The first job was to measure the circumference accurately. They knew nothing of 2πr [8], but they had a very infallible method. This vital piece of equipment was a wheel with a handle attached. This wheel was run round the circumference of the wheel and the number of complete revolutions counted, and the extra bit between revolutions was marked with a piece of chalk. Very crude you might say, but road markings are measured in a similar, but more sophisticated, way today. Next, the bar of metal for the hoop was laid flat and the wheel was run along the length the desired number of turns.

---

[8] 2 Pi r is the formula for calculating the circumference of a circle. Pi is 22 divided by 7 (total is 3.1412), r is the radius and 2r is the diameter

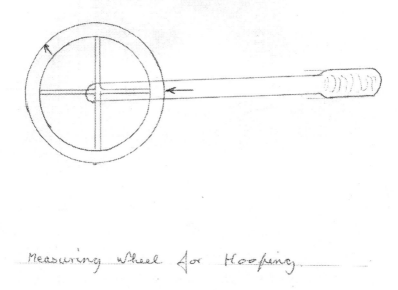

Measuring wheel for hooping

This was marked out and the straight length measured, usually with a 2 ft rule. Having ascertained this measurement, ⅛" was deducted for every 12" of bar, and that was marked on the bar to be sawn off at that point (⅛" for every 12" being the length that the bar would expand when hot). When the hoop was cooled it would shrink back to its original size and tighten all the joints and cling to the wood.

Having cut our bar of metal, (which had to be wrought iron to make a good weld), to the required length, it was now put through a rolling machine (a bit like a mangle) and formed into a circle. The two ends would then be heated to a white heat and hammered together with the help of 2 strikers, wielding sledge hammers, thus forming a welded joint. This 'fire weld' was very critical because if it was not done properly it would break when the metal cooled and the shrinkage put enormous pressure on it.

To form the fire for heating the whole ring we worked outside the middle shop, where we had a round trench about 6" deep, and across the top were plates with holes in. This was connected up to the fan inside and air was blown through.

The two big hearths inside were stoked up high with coke until there were two roaring fires. Then the hot cokes were taken outside on shovels and placed onto the plates on top of the trench, and a ring of fire was soon roaring away. Three men with long forked handles would then lift the hoop and place it in the ring of fire, turning it occasionally to get an even heat all round. While this was going on another two men were preparing a big, flat, cast iron ring with a hole in the middle, set up on 3 stone blocks. The wheel was now placed on the flat plate, with a hook fixed through the hub. This was for lifting the wheel and hoop and plunging it into water. Three buckets of water were placed round the ring, ready to pour onto the wood if there was some delay causing it to set on fire.

As soon as the metal ring was hot, three men with forked rods picked it up and placed it on the wheel. It was then hammered down level onto the supporting plate as quickly as possible, and before it could burn any wood away it was lifted by the crane and dropped into a well of water until it was cold.

That was hooping.

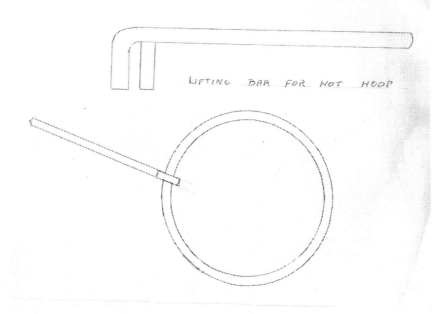

Lifting bar for hot hoop

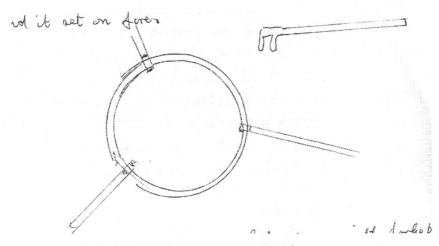

Lifting of hoop required 3 lifters

Brick Nicker

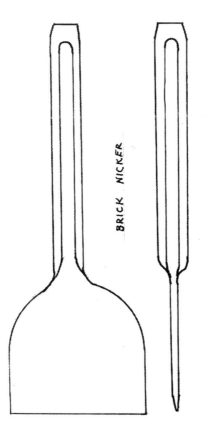

The brick nicker was different from the stone nicker as the blade was flat and thin with a very sharp end. Usually it had 7/8" octagon handle and 3" wide blade. A smaller version of this was of similar shape and used by joiners for cutting the tongue off a floorboard when cutting into a tongued and grooved floor.

Chapter 6

The Factory Inspector

We were aware that all the machines did not have guards which complied with all the factory regulations, but we thought if we were careful we could take a few liberties. The saw was a machine which ran from a belt from the main shafting, the belt coming down at an angle of about 60 degrees. When the belt got a bit slack and needed taking up – or shortening – or if the saw blade was worn and jammed, the belt was liable to come off the pulley. We had a method of putting it back on without stopping the line shaft. It consisted of a wooden pole about 6 feet long with a little angle bracket 2" from the top. This could be hooked under the belt, and the belt could then be guided back onto the pulley.

Once you knew how to do it, it would take about 2 minutes. You could not do this with the guard in place, so you had to stop the line shaft (and every other machine in the whole works), get a ladder and ease the belt onto the pulley. There was just one snag. As you walked through the workshop, you had to bend your head and shoulders sideways at the appropriate place to miss the belt, but it was no problem as everyone had this manoeuvre off to a tee.

Another machine which was not to standard was the grinding machine, or emery wheel as we always called it. We did not have a guard to the side, so that we could grind our chisels and hammer faces on the side.

Father, being aware of these inadequacies, had instructed me thus – that if the factory inspector came, he would keep him talking in the blacksmiths' shop while I organised a quick assembly of the guards. Unfortunately we did not discuss a communication system.

One pleasant afternoon the sun was shining, and as I walked from the welding shop towards the blacksmiths' shop I was suddenly confronted by Father's bowler hat rolling out of the door on its brim. The factory inspector had arrived and was inspecting the machine shop at Father's side so he was unable to do the shoulder bend and the saw belt had whipped his hat off. The inspector was not pleased and made many 'suggestions' that he would like to see put into place next time.

Holeing Pick and Splitting Wedges

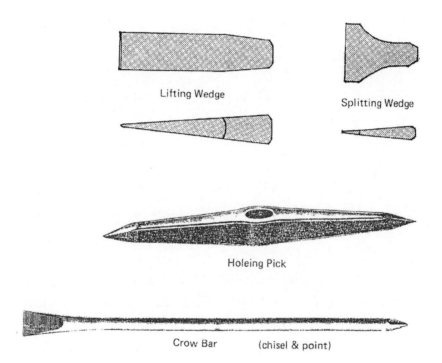

The holeing pick was used to cut a groove in the stone. It was about 18" long and forged from 1½" square steel. It had a taper shaft for easy removal when the pick needed sharpening. When the groove was made to the depth of the tapered end another smaller pick of the same shape was used to open out the bottom of the groove. This was known as the bottoming pick. The stone was then ready to receive the splitting wedges which were placed at intervals in the groove and driven in, in turn, using the 14lb sledge hammers. The stone was thus split into manageable blocks. The pick has long since been superseded by the electrical hand-held abrasive cutter.

Chapter 7

Rastrick Illuminations

At the time I was learning the trade, we did not own our own property. The properties on one side of Ogden Lane, from the Grammar School at the top to the smithy at the bottom, belonged to a very old charity named the 'Mary Law Charity'. It was administered by the governors of the Grammar School to whom we paid rent. Our premises always looked shabby to say the least; and although we whitewashed inside every year, it was an event to paint the outside.

Every year we had holiday on Easter Monday and Tuesday; we did not actually work on Good Friday, but we did not rest. It was a way of observing Good Friday by not doing our usual job and using the hammers, but at the same time making use of the day by cleaning, 'siding up'[9] and whitewashing the walls.

Every man was given an old sugar sack[10], which then had holes cut in for his head and arms; this you donned for your protection. First of all, everything hanging on the walls was put outside, tongs, swages, tools, hammers, the lot. Now, armed with protective sack and a long brush, we proceeded to sweep down the walls and rafters and dispose of all the muck[11], if we hadn't choked already.

---

[9] Putting things away
[10] At the time sugar was delivered to local grocers in Hessian sacks and then weighed out in bags to customers requirements. The grocers were glad to get rid of them for 4d each.
[11] dirt

While the walls were being brushed clean the apprentice would have been dispatched with a wheelbarrow, to Harry Castles Builders at the bottom of Huddersfield Road, near the end of Lords Lane. Here he would pick up a barrow load of white lime and a bag of 'dolly' blue powder. The lime was very soft, you could shovel it, but when you got a barrow full it sloshed about and was very heavy to push back up Rastrick Common to the smithy. Even though lime was white, if you did not mix some blue with it, it would dry grey.

When we had cleaned the walls, a quick dip of the brush head in water to remove all the debris and then on with the white lime; there was no time to let it dry. Everything was put back in place and, hopefully, by about half past three in the afternoon we could stand back and admire our handiwork and have an early finish. Sorry, I digress a bit here; we were talking about the shabby exterior.

One day some painters arrived in Ogden Lane and, starting at Mrs. Parrat's shop, proceeded down the hill towards the smithy. Although my father had not been informed officially, all the other tenants said that the charity had decided to give the properties a facelift.

We awaited our turn, but it never came! The painters stopped when they reached the smithy.

A couple of telephone calls later, it transpired that we were not included in the scheme, being industrial premises. The man was not very charitable and said,

> "If you want it doing, you must do it yourselves."

This displeased Father greatly. When he enquired if we could choose our own colour the man from the charity said, "of course".

Although the place hadn't been painted for years, it suddenly became a priority.

> "Catch the next tram to Brighouse", ordered Father, "And bring a can of every bright colour of paint you can find".

I went to John Francis Brown's , where we had an account, and chose a tin each of red, blue, green, yellow and pink and some brushes. The following day I became a painter. Each rail in a row was painted in a different colour in turn. The window frames the same; every spar different, the doors and the crane also! People in the street stopped in their tracks and enquired if this was some joke. I assured them that it was deadly serious.

Word spread around and the following Friday there was an article in the Brighouse Echo entitled 'Rastrick Illuminations'. On hearing about the episode, my father's aunt, Annie Whiteley (maiden name) who was a close relative of Uncle George (George Shaw Whiteley)[12], wrote a poem of many verses, describing the whole affair. Unfortunately no copies of that are available, but the last verse read,

> "This little disparity,
> Which has caused much hilarity,
> Is not very much to relate,
> For it all is the property
> Of Mary Law Charity,
> And helps to keep up the estate."

---

[12] The founder of the firm.

Plugs and Feathers

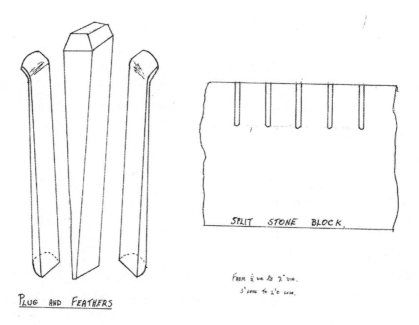

These were for splitting stone of all sizes. They were made to fit a particular size of hole and the length depended upon the diameter of the drilled hole. A set for a 2" hole would be about 18" long; while a set for a ¾" hole would be about 6" long. When fitted together the plug and feathers fitted the hole for the whole depth. This put pressure down the whole length when hit with a hammer. The wedges were hit in turn until a split was formed right through the stone. If a wedge was used without the feathers it would simply break off pieces at the top and not split through.

Chapter 8

Local Transport

It seems strange, looking back, to think how we put up with the methods that were available to us at the time. As blacksmiths we had a metal barrow, the lads from the local quarry had a wooden one, while the local joiner would have a flat cart with two wheels and a long handle, on which he pushed his timber or finished article.

The local quarries sent a boy with a barrow full of tools to us to be sharpened and he took a load of sharpened ones back, but otherwise we had to deliver. At the end of Birds Royd, adjacent to the railway line, were what was known as the 2 "stands", the stone stand and the coal stand. Coal was delivered by rail to the coal stand and the "Coalman" would collect his coal from here with a horse and cart and then deliver it locally. The opposite applied to the stone stand. Stone was delivered on big wagons to the stone stand, and then the stone mason would dress it to the required shape. It was here that stone flags were dressed and then rolled end over end onto railway wagons to be sent all over the country.

It was our job to keep them supplied with sharp chisels and that meant the apprentice pushing a barrow full of sharpened chisels down to the end of Birds Royd and pushing a load of blunted tools back. This was no mean feat when you consider that Rastrick Common was paved with stone cobbles, with tram lines running through them, and Birds Royd was paved with granite setts; and all for 10/-[13] a week.

It was not convenient to use this method for the quarries in Southowram and Lower Edge. They had their own little blacksmiths' shops and we used to send a man for a day or two days a week to sharpen the chisels on the spot. The method of charging was by a points system. A punch (up to ½ inch wide) and a 1 inch chisel counted 1 point. A splitting chisel (nicker), at 3 inches wide, was 2 points and you were expected to do those at 1 heat. A thicker pitching tool, or pitch nicker[14], as it was often called, had to be heated twice, once to hammer to a level edge and once to file the edge while hot, with a rough file. This was 3 points. A pick was 2 points. The number of points for each particular batch of sharpening was added up and then charged at, say, 2/8 per score.[15]

When we made new tools for the quarries in Southowram we used to send them by bus. At that time the bus station was in Thornton Square and Halifax buses left every half hour. Next door to the Black Bull public house and in between Pollard Ives' grocers shop there was a little cubby hole with one door where parcels were stored to put on the bus. A young boy employed by Halifax Corporation was in charge and he would load and unload them onto the buses, but not before they had a sticker on the label, which had to be obtained at the grocers. Mr Ives would lift the sack with one hand, as if weighing it and calculating the cost, and then say 4 pence (old money); he would then apply the required sticker. A telephone call was then made to the quarry and they would send someone to meet the bus at the Delvers Arms public house.

---

[13] Ten shillings
[14] A pitch nicker was for forming the rounded face on building stone, see illustrations.
[15] Two shillings and eight pence per 20.

Sending goods to Huddersfield was by tram. We used to send cold chisels to Gregory and Sutcliffe, engineers suppliers there.[16] Harry Mayhall at the newsagents shop in Church Street at the top of Ogden Lane was the local agent. He would apply a 4d sticker to the label, and then you had to hand it to the conductor on the next tram. This would be put behind the step to the upper deck and then deposited at the goods office at the top of Northumberland Street. From here it was collected by the customer. It is interesting to note that on the back of Huddersfield trams there was a post box and you could post a letter up to 8.0' clock at night.

The railway goods yard in Brighouse was one of the largest in the area, with rows and rows of sidings, where trains were marshalled all night. There were large warehouses too and goods were collected and delivered by horse and cart; the railway company had 20 horses before the war. They were very flexible and we could have a parcel collected in the late afternoon and delivered to, say, Baliff Bridge next morning. Goods were collected from the mills in Rastrick every day, so all we had to do was hang a large sign on the railings, 'LMS To Call', and the wagon stopped and collected.

Towards the end of the war, as the system was becoming mechanized and the horses pensioned off, the first trucks were called mechanical horses. The cab was mounted on three wheels with a small wheel at the front. The cab was then connected to a flat wagon. You could have more wagons than cabs so that a wagon could be loaded while one was being delivered.

---

[16] Cold chisels were for engineers for use on metal and very different from masons chisels.

Lifting Lewis

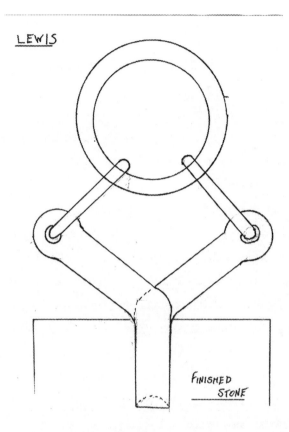

When a piece of stone for a new prestigious building has been sawn and polished it needs to be lifted without damage which might be caused by lifting with a chain or other method. A simple method is to drill a hole in the top of the stone and insert the half round ends of the lewis. As the weight is lifted the legs of the lewis tighten and the stone is handled with safety.

Chapter 9

The shearing machine

The shearing machine for cutting bars to length was not one that you could stop and start at once. It relied on the momentum of a 6' diameter cast iron fly wheel and the jaws opened and shut at a fairly fast rate. We could set a gauge behind the jaw and after each cut you had to move the bar forward before the jaw came up again. As it was belt-driven, to stop one had to guide the belt onto a free running pulley and allow the fly wheel to lose momentum and stop in its own time. We had a guard to put over the blade so that the machine could stop safely. Sometimes when cutting a lot of lengths one might jam and then it was impossible to push the bar through for the next cut. To overcome this we had a hook and handle with which to pull out the offending piece. This could be accomplished without danger. Everyone was told that they must use the handle at all times to clear a blockage.

Of course there is always the one time when for some reason or other someone does not do what they have been specifically told to do. On one occasion, when cutting some lengths for the Riley link, there was a blockage as one length jammed between the lower blade and the length gauge. Without thinking one young man, instead of using the handle, reached in behind to clear the obstruction and he put his index finger between the blades. Of course off came his finger. The factory inspector was at the door in no time at all. He went through the machine shop with a fine toothcomb, and required us to alter every guard on every machine except the shearing machine – the one which had caused the accident! He could not find fault with that.

Of course we were sued by the lad concerned. In court, his solicitor said that his client was working for us temporarily as it was his ambition to join the army later on. By losing the trigger finger of his right hand he had lost his career.

The incident cost us a lot of money.

Dummy Dog, Saw Frame Dog and Cotter

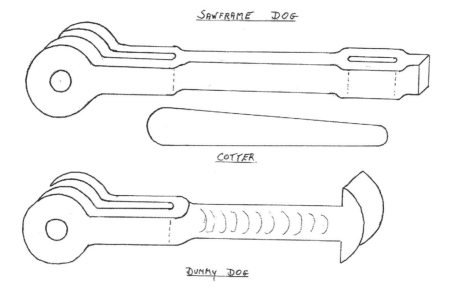

Large blocks of stone (Jud[17] lumps) were taken to the saw yard for sawing into flagstones for pavements. The saw blades were about 10ft long and 6" deep and about ⅜" thick. They fitted into a frame with a swinging motion and were made tight by driving the taper cotter into the slotted ends of the dogs. The dummy dog had to be used where the connecting rod fastened to the saw frame. Water and steel shot was constantly poured over the stone and the shot dropped between the grooves and onto the stone. It was a wearing-away process and the saws ran day and night and only stopped to load another block.

---

[17] Random lumps of stone as they come off the stone face.

Chapter 10

Siege Bands

Before I started work, whilst still at school, an accident happened with the old metal shearing machine and it was fortunate that no one was seriously hurt. The 6ft fly wheel was going at full speed and chopping lengths of steel one every few seconds. One had to be very quick to push the bar between the jaws to the stop, as they rose and fell methodically, being geared to the fly wheel.

On this occasion, instead of a bar of mild steel being sheared, a bar of very high quality, hard tool steel had by some misfortune got mixed with the mild steel. The resulting bang, like the sound of a cannon, and the spectacle of the fly wheel coming to a dead stop and the top of the machine falling to the floor as the casting broke in two all happened in the fraction of a second. This was a catastrophe at the time, and the broken casting weighing about half a ton was laid to rest in the yard until it was decided what to do with it. It was about 20 years later when the answer came. We received a bundle of blueprints from a firm in Birmingham, for some siege bands and tie rods for use when building glass furnaces (at that time there was no copying or taking prints off as we know it today. The draughtsman's drawings were photocopied and came out as a white line of blue paper). Siege bands were made from 5" wide * ⅜" thick material, and were bent to a specified radius. On the end of each was riveted an angle bracket made from 5" wide * 1 ¼" thick steel. These were then bolted together to secure the brickwork of the furnace when it was hot.

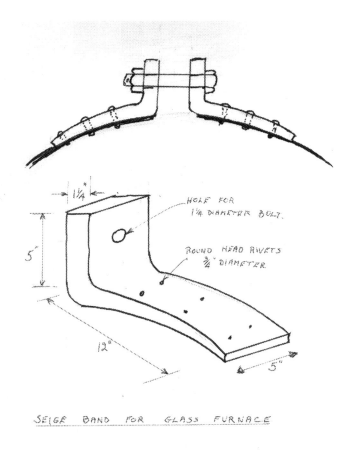

Siege Band for glass furnace

These angle brackets weighed about 60 lbs each and had to be handled with a pair of tongs. The problem was how to hold them to bend them. It was impossible to bend them over the anvil. Then the idea came; the old shearing machine. It was robust, heavy, bedded into the ground after many years, and would not move. We were able to fasten some heavy brackets to the old casting and then drop the hot bar in a slot to the required depth and then bend the top over. The idea worked well and was to be used for many years. It took 3 men to bend these heavy brackets. We worked as a team. Number 1 handled the flat bar and heated it in the fire and carried it out into the yard, and put it in the slot. The other 2 men (numbers 2 and 3) were waiting with sledge hammers (14 lbs heavy) and alternately struck the end over and flattened it level. It was then put aside to cool. Next, number 2 heated the metal while number 1 picked up the sledge hammer. Following that, number 3 heated the metal while the others did the striking, and this rotation continued until they were all bent. The next process was to taper the long end and bend it to the required curve. A piece of flat metal was placed on the floor and, using a piece of string cut to the right length, a line was marked in chalk showing the correct curve.

The blacksmith then placed the bracket on the drawing and hammered it until it fitted. Two men using 2 fires would do this. When they were cold, the 3rd man would then mark them out and drill the necessary holes for the rivets. The next job was to bend the bands. For this we had a rolling machine worked by hand. It was like a mangle but with 2 rollers at the bottom, and a moveable one set between and above the other two. Having bent these, they had to be accurately marked out and drilled so that they matched exactly the holes in the brackets. Riveting was again a 3 man job. There were 6 rivets to put in, but firstly the brackets were bolted in position with 2 bolts. It was then easy to put 4 rivets in, remove the bolts and complete the job. One man worked the crane, and putting the hook in one end he lifted it to the right height. Another man sat astride the band and manoeuvred it so that the rivet came through the hole. We had an anvil out in the yard, on the ground, and in the anvil was a little cup to take the rounded head off the rivet. The 3rd man was inside getting the rivet hot. When it was white hot he brought it out with a pair of tongs, placed the head in the cup and held it till the bracket fell into place. He then picked up the sledge hammer and flattened the rivet while it was still hot.

This might seem a tedious process but it did work well for a period of years. It was many years later, when we acquired a 50 ton press and a gas furnace for another job, that we realised how much time and energy had been spent bending those brackets. We were able to put 6 plates in the furnace, wait an hour for them to warm up and then bring them out and bend them under the press, about 3 minutes each. That was progress.

One of the jobs that we did for our customer in Birmingham was for the Fountain Glass works at Gilderstone. While construction of a new furnace was nearing completion, it was found that some bolts has been misplaced so we had to make some more. I delivered these one day and was privileged to have a look round. I was very impressed with the big machine reaching nearly to the roof and as wide as the building. Molten glass was going in at the top, and out at the bottom emerged, in their hundreds, glass jars for holding Ponds cold cream.

However, some 12 months later, that shed was quiet. There was no machinery working, no hot glass dropping in the top; everything was cold and dusty. In their wisdom Ponds' had decided to change from glass jars to plastic ones for their cold cream. That was progress too.

Wire Rope End Socket

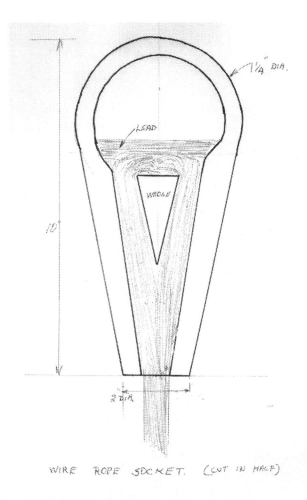

WIRE ROPE SOCKET. (CUT IN HALF)

Sometimes it was necessary to either change the socket end of a wire rope or fit a new end on. A piece of wrought iron, about 2" diameter, was split at the end and the two ends forged round. They were then formed into a ring and welded together. The hole was then drilled down the centre and a taper mandrel was hammered into the hole. The wire rope was then inserted into the hole and a round wedge was driven into the centre of the rope and the ends bent over the top of the wedge. The cavity was then filled with molten lead. The harder it was pulled, the tighter it became.

Chapter 11

Pointing Irons

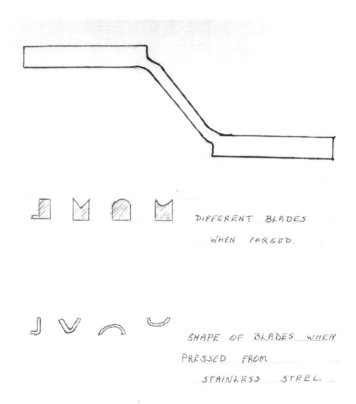

A pointing iron was a tool unique to this part of the country because of the abundance of stone for building. In the southern part of the country brick was more popular for building and the seams between the bricks were very standard. They would either be smoothed off with a trowel or, more often, with an old bucket handle, which left a smart convex seam.

With the stone trade it was different. One had the choice of a convex, concave, V shape or flat type of seam.

The flat-face pointing tool was about ½" or ⅝" wide with a lip ⅛" deep at the edge. The face of the pointing iron had to be smooth. If it was rough or rusted the mortar stuck to it and the smooth finish was ruined. The flat type was used for the larger seams of random stone walling and was pulled across the seam at a slight angle. The lip at the bottom cut the mortar off clean. When it rained the water was drained away by the slope of the seam and then it dripped away because of the clean cut off at the bottom. This prevented water lodging in the seams.

When a job was finished the pointing iron was often dropped into the tool box and forgotten until the next job, by which time it would be rusty and it would then have to be shone up again before it could be used.

One day a rep from the local builders' suppliers came into the works and dropped a stainless steel pointing iron onto the bench.

> "Could you make these?" he said

I looked at the sample and immediately saw the possibilities of this tool which would not rust.

> "Of course we can make it", I said

> "Drop us a price in the post, then, and we will see if we can do business."

Clifford Riley

I got in touch with the suppliers of stainless steel and estimated our costs. The stainless steel was a good idea, but a bit expensive. I submitted a quotation. Later I received a telephone call which gave me a shock. It transpired that the sample was made in Japan and was a fraction of the cost of my quote. The finished article all the way from Japan was cheaper than the cost of the material that I could buy from our suppliers. Fortunately for me this sample only came in one pattern, which was convex. This made a concave seam, just like a bucket handle. However, all was not lost. I pinched the idea of using stainless steel. I made both concave and convex patterns as well as V shaped and flat. First of all I made various tools to form them under the 50 ton press, It was then easy to form the different shapes from ⅛ inch thick stainless steel plate, which made them light and rustless. They were much easier to make than our steel type, so we were able to run the two types together at a similar price.

Rock Dog and Nip Dog

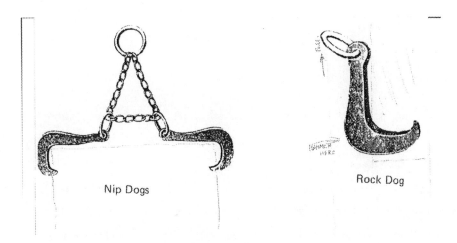

Nip Dogs

Rock Dog

The rock dog was much bigger and heavier than the nip dog. It was hammered into a seam and used for dragging blocks from the stone face, to be removed using nip dogs.

When a large block of stone was being lifted by crane 2 dimples were chiselled at the appropriate distance from the edge and the points of the nip dogs were inserted. As the crane was raised the dogs were pulled closer together and the stone was held fast.

Chapter 12

The weather vane

It gives one a feeling of pride when asked to do a job that is there for every-one to see, hopefully for many years to come. I had this feeling when the vicar of St Matthews at Rastrick came to see me one day, with a piece of rusty old wrought ironwork in his hand. He explained that it was part of the Church weather vane, which was unsafe, and that the whole thing had to be taken down for safety. The church elders had met and they felt that, for the sake of tradition, they would like a new weather vane making exactly like the old one. Could I – and would I – undertake to do this job? He assured me that the steeplejacks, who were doing remedial work on the dome, would erect it. I was delighted to say yes.

The steeplejacks brought all the old bits and pieces to the works and we set them all out, and took all the necessary measurements. The central shaft was 20ft long – 2" in diameter for 12 ft, then tapering to 1 ¾" in diameter then down to 1 ¼" in diameter. The letters were 12" square and were cut from ¼" thick plate. The cross pieces to which they were welded were 8 ft long. There was a 3" diameter boss in the middle and this was bored out and a brass bush was pressed into the centre to take the spindle of the moving arm. This was so that the two surfaces would not rust together. The pointer was in one long piece, and the opposite end to the point was encased between two pieces of copper sheet, 3 ft long and 12" deep (this being to catch the wind).

When it was completed, the pointer was balanced. Two metal discs of the exact same weight were welded to the point to make each end of even weight, and to ensure as little friction on the bearing as possible. The finished scrolls were 3 ft and 2 ft 6" long, and made from 1 ½" wide and ½" thick mild steel. Round collars were then made to slip over the top of the centre pole, and fastened in position with bolts. The scrolls were drilled and then bolted to the collars. In this way they could be adjusted to fit exactly where required. A steel girder was now prepared and this had to be let into the sides at the bottom of the dome, and fixed centrally. A socket was then made to hold the bottom of the central pole, and this was made to slide on the girder. When this was in the exact position for the pole to be vertical, it was fastened with bolts. Lastly the lightening conductor needed to be fixed to the top. The original lightening conductor was made of copper and was in good condition, so this was cleaned up and placed on top of the 20 ft central pillar, making it 26 ft high. The whole weight was approximately 600lbs.

When the job was completed my father suggested that we erect it in the yard for everyone to see, and to be aware of how big it really was. This was much appreciated and the Echo sent a photographer to take pictures. For a long time, every time I passed the Church, I would pause to have a look at my handy work. I was very pleased when one day I saw a crow land on the pointer and it moved round. I knew then it was going to be ok.

Clifford Riley

Rastrick
Church's new weather vane weighs nearly 5½cwt.

WORK commenced on Tuesday on the fixing of a new weather vane and lightning conductor on the top of the tower of St. Matthew's Church, Rastrick. Some 26ft. high, the vane is a replica of the old one, believed to be over 100 years old, which was removed a few months ago after it had become dangerous because of its corroded condition.

For a few days before the job of fixing it to the church dome was started, the new vane was inspected with interest by many people as it stood in the blacksmiths' yard of G. S. Whiteley and Co. at the bottom of Ogden Lane. The arms carrying the letters indicating the four points of the compass are 8ft. across and the letters, painted gold, are themselves 12in. square. The wind-indicating arrow is 7ft. long, the main shaft is of 2in. tempered steel, the total weight is about 600lb, and Mr. Clifford Riley, only son of Mr. John Riley, head of the firm, has spent three to four weeks on his task, endeavouring to make the vane an exact replica of the old one.

The vane was dismantled again before being hauled to the top of the church tower, and there it is being reassembled and erected by a Bradford firm of steeplejacks. A steel girder has been erected inside the dome, in place of an old wooden beam which had supported the base of the old vane, and the new shaft fits into a socket fastened to the girder. The north position is being set by compass before the arms are locked in position.

Fitting the new weather vane

Lifting Wedges

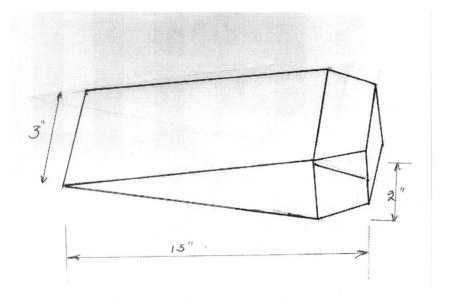

Lifting wedges were heavy wedges used in the quarry for driving into a seam on the face of the stone and lifting it. Crowbars or rollers could then be used with chains slotted underneath, to lift the blocks.

Chapter 13

Sladdins

Just above Rastrick Church on the right was Clays Mill, where some of the finest cloth in the country was woven and exported to America. I remember Travis Clay, who was an Alderman and governor of the Grammar School. I can see him now. A big man in knee breeches and a yellow waistcoat as he used to stride into the school to see what was happening. Sadly, as the woollen trade disappeared the mill had to close.

It was taken over by Sladdins, who made shoulder pads for suitings. Rags were fed into 2 large rollers running together. The rollers were covered with what was known as carding. Carding was a flexible belt through which many wires had been inserted (like a wire brush for brushing suede). All these thousands of little teeth tore the rags to shreds which finished up as fluff. As the fluff came out of the machine, it was pressed and then passed over a roller which was dipping in glue. At the end of the process the felt came out like a long belt. Shapes were then stamped out and the end product was shoulder pads.

All the machinery in the mill was very old and we were often called upon to weld broken castings and cog wheels. With the cog wheels we had to build up new teeth and file them to the correct contour. Sometimes we had to build up shafts and fit new bearings. The engineer who used to bring the casting to repair always had fluff on his eyebrows and cap.

We knew him as 'slack fit', because whenever we renewed a bearing he would always say,

> "Make it a slack fit lad; I don't want it to overheat if it gets short of oil".

With the hot glue and fluff it is a wonder the place was never burned down.

Claw Tool Holder

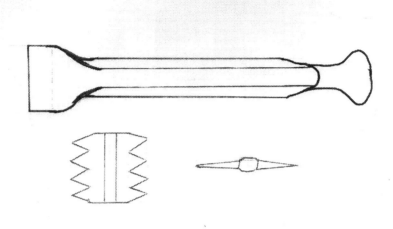

CLAW TOOL HOLDER WITH SLOTTED END.

This tool was used mostly by Monumental Masons who had no access to a Blacksmith. The claw bit was made from hardened steel. When it was blunt it could be turned round and used at the other end. The head was smooth and rounded for use with a wooden mallet.

Chapter 14

The Riley Elevator Truck

One day a stranger drew up at the gate and hesitantly came down the yard. He carried a small attaché case and asked to see the manager – as Father was standing nearby in his bowler hat I motioned him down the yard. After a while I was called to join them as they browsed over a little wooden model set up on the bench. It consisted of about 6 slats of wood nailed together thus:

THE "RILEY"
'ELEVATOR'
HYDRAULIC WARP TRUCK

Patent Nos. 619431 and 36409

(Type B)

*

Enables one man to put a warp into a loom. Suitable for lifting skips and bales. Ideal for negotiating the narrow alley ways between looms, easy to manoeuvre. Can be turned completely round in its own length, and be lowered slowly under control to any desired height. Thoroughly tested before delivery. The rear wheel is of the castor type giving extra manoeuvrability while the table is offset 2" to facilitate the approach to the loom. All our trucks are felt covered and fitted with rubber wheels or castors. If your beams are to load and unload from, and on to the floor, we can supply a ramp for rolling these on to the truck.

Height when elevated	2ft. 8in.
Height lowest position	11in.
Wheel base length	3ft. 0in.
Wheel base width	1ft. 6in.
Table	1ft. 8in. x 2ft. 6in.
Working load up to	850lbs.

The man was holding the contraption at the top and lifting it up and down so that the arms folded in a scissors movement. He then turned to me and informed me that he was a supplier of goods to the cotton mills in Lancashire and he had been asked to supply a device to put a beam of cotton into a loom. It had to fold low for loading, be narrow enough to wheel between the looms, rise above the existing beams in other looms, and then lower the beam into the sockets of the loom. Rows of looms were set back to back and there was only 8" between the beams. As the weaving proceeded and the cotton was used up, the remaining spindle could be moved by hand. Could we design and make such a device?

I thought of all those looms in Lancashire, and thought the scope was enormous. I said I would draw something out and would be in contact. At that moment I didn't have a clue about how I was going to do it. I talked it over with my father and we tossed a few ideas to and fro and decided it would have to be hydraulically operated from a pump, with a lever to be pressed by foot. I got my drawing board out and set to work. The scissor movement was no problem, but I was no expert on hydraulics so had an idea to use a car jack. I was delighted when we did a mock up in the workshop, and the idea worked except for one thing. There was no room to work the car jack when the beam was loaded. It needed a separate ram, and an oil pump in a position that the operator could use it. I drew this out on paper, but the next thing was to make it.

The making of the basic pump was straightforward, but I then had to find suppliers of all the bits and pieces. I needed some gland packing to stop the leaks, brass nuts to tighten the packing, ball bearings to make ball valves, hydraulic tube for the ram, cup washes for the end of the ram and pump spindle, wire mesh to make sieves to keep out the fluff, copper tube to transfer the oil. The list seemed endless, and that is only the pump: now for the truck itself. We needed hard rubber wheels and castors, slotted nuts to stop them coming loose and felt to protect the top.

We seemed to be spending a fortune, and we still hadn't got one to show. Eventually, however, the truck took shape and we submitted our prototype and waited with bated breath. Apart from a few little adjustments, we were set to start production. The truck sold well and we were so pleased that we started to think about a larger model for the Yorkshire woollen mills. The beams were much heavier and longer, but still needed a fairly narrow middle aisle and this we were able to achieve.

The next job was patent rights. We didn't see how 6 sticks of timber could be patented, but enquired about the possibilities of patenting ours. What a job that was! I had to do not only an assembly drawing, but a drawing and minute detail of every single piece that went to make up the whole. The patent was granted and we were in business.

They say that all good things come to an end, and the inevitable happened to our trucks. After the war, new and better looms were being produced, looms were re-spaced to give more room for workers, and beams became bigger and heavier and fitted near to the ground so our truck would not go low enough. It was alright while it lasted and we made in the region of 300.

Chapter 15

Wrought Ironwork

Thornhills Lane, Clifton

I was always interested in wrought ironwork, and compiled a catalogue of designs for gates and door grills and made many in my time. There were 2 parts of wrought ironwork: the sort the blacksmith made where all the ends were tapered and the scrolls hand-forged; and the other sort which could be bought through the adverts in the Sunday paper, cost half the price and were mass-produced out of thinner material. We could never compete on price but we could on quality. Flower stands and candle holders were another line and examples of these may be seen in Central Methodist Church and St Chad's Church.

Flower arranging stand, Central Methodist Church, Brighouse

Another job was to make the railings round the children's corner at former Park Church. These had to match the ones round the communion rail. I also made a 6-sided lantern, with a 500 watt bulb to hang over the pulpit, to help the minister or preacher to see their notes or read the bible. The minister came to work to see how it was progressing. It was almost finished and he enquired how it was to be held up. I showed him the chain. He looked at the lantern, held it in his hand, looked at the chain, then looked at me and said:

> 'If you think I am going to stand and preach with that over my head you must be mad'.

I answered that it would be quite safe, and lifted a weight four times heavier than the lantern on the chain to be used. This convinced him and he accepted that it was a boon. When the minister left I was asked to make him a wrought iron lamp standard as a going away present. Instead of the usual bulb and lamp shade, I made an enclosed lantern for the top and this was much appreciated.

Table made in 1950, for my new house, when we got married

The best gates I made were for Mr Brooke of Brooke Chemicals in Hipperholme. He was moving house and was loathe to leave the gates at his old home, because when he was younger he designed them himself and had them made in London. However, the gate posts of his new home were 30" wider, so he had to leave them. He decided to have a pair of gates made to look exactly like the original ones, but fit the new gate posts. He was discussing his dilemma with the builder (who happened to be a good customer of ours) and considering some London firm again. The builder assured him that there was no need for this, as he knew who would make them. The man was dubious but agreed to see me. He said that if I could show him a drawing, showing how I would account for the extra width, we would talk about making them. I very carefully measured every piece, every bar, leaf, scroll, and catch … everything.

A week later I submitted my drawing. He was very surprised, agreeably surprised. He studied the drawing and then looked at me over the top of his glasses and said, 'If your work is as good as your drawing, the job's yours'.

I got such a lot of pleasure making those gates. I made the scrolls and leaves and Joe welded the frames together. They weighed over half a ton and I was very proud when they were erected and Mr Brooke was full of praise. He had had no idea that there was anyone locally who could do such a job. Some time later, the cheque for the gates came through the post, and Mr Brooke died the following day. The gates still look as good as ever at Lightcliffe.

Clifford Riley

The wrought iron gates for Mr Brooke and Joe Cunnliffe, the welder

Chapter 16

Odd Jobs

Nora, along with some of the ladies of Clifton Church, was a member of a Luncheon Club at Cleckheaton. I am not sure what ladies discuss at these events, but I am sure that husbands and work come up from time to time. It was after such a meeting that Nora said over tea,

"Could you make a treasure chest?!"

I thought she had come into some money, but no such luck. In conversation with her neighbours at lunch it was mentioned that we had a small blacksmith and engineering shop.

"You might just be able to help me", enthused one neighbour.

She then explained that she and her husband always bought unusual gifts for each other on their birthdays. It transpired that the husband was a compulsive biscuit eater and that he had his own supply of a special brand. She had the idea of a small treasure chest with lock and key, so that she could ration his biscuits; whether or not he would appreciate the surprise, seemed of no consequence. Did Nora think I could make such a thing? - price was no object.

Nora was sure I could, and a meeting was arranged. We discussed sizes, the lady stressed she wanted a good job and that it must look the part, quite old. I had never been asked to make a new job look old before.

I made it with ⅛" steel plate with brass bands over the top and sides and a loop staple for a tiny padlock. I was very pleased with the job so far, but it looked perfectly new. I contacted Vernon Moss on Churchfield Drive. His firm did a lot of nickel and cadmin plating for us for our ICI jobs. Could he put an antique finish on a small metal chest?

"Bring it in", he said.

He was intrigued by what I was doing and entered into the spirit of the idea. What he did, I know not, but the finish was perfect. To finish it off, we fixed a brass plate on the chest with the man's name and date of birthday engraved on it, wrapped the chest in tissue paper and waited for it to be collected. It was opened with sheer joy and admiration and I felt a flush of pride at a job well done.

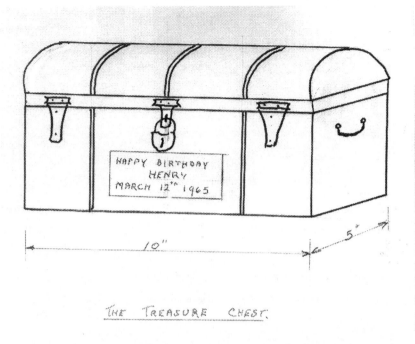

The Treasure Chest, a wife's present to her husband who was a compulsive biscuit eater

One morning a local trumpet player came down the yard in great haste.

"Can you do a lathe job for me, urgent?"

"How soon is urgent?" I asked.

"Now", came the answer, "I have a *Messiah* at St Peters' Church tomorrow, and we had a rehearsal last night and I couldn't tune to the organ. I've rung round and it is impossible to get a suitable new mouthpiece. I have an old one here and I want a few thous[18] turning off the tapered end to enter my instrument a fraction further".

"Look at this lathe", I said, "it's 30 years old, it turns rough forgings, it's more acquainted with 1/16^{ths} than thous".

"Stop messing and get on with it, for goodness' sake"

(Being my cousin, he could talk to me thus). With great care I managed to get it running true, and not daring to use a turning tool I used a new, very smooth file and gradually managed to file it as it turned in the lathe. From time to time we stopped while he inserted the mouthpiece and gave it a blow.

"A bit more", he said.

"What happens if I take too much off?" I enquired.

"Don't think about it", he replied.

After three attempts he blew a resounding note, and a beam of satisfaction creased his face.

"That's it", he said, "smack on",

And with profound thanks he was on his way to the Church to try again to tune to the organ. *The Messiah* was a huge success, and received rave notices in the press, I am happy to say.

---

[18] thousandths of an inch

A surprise challenge occurred when a few men were chatting together, waiting for their wives, after a concert. The minister was explaining that he had a problem. The church was hosting a very big ecumenical event in two weeks' time, and during the course of this it was arranged that the congregation would take communion. It would be ceremoniously performed and the wine would be poured into tiny glasses or silver cups and then distributed to the people for them all to partake together. They would need four servers with the usual silver jugs, but had only two jugs which were in normal use and shiny, and they would need to use the other two which had been stored away for many years. These were black. The minister had contacted all the local jewellers to see if they could finish these two without scratching them. No-one wanted to tackle the job and all explained the various difficulties. I had to put him out of his misery.

"Let me have them", I said, "I'll do them for you".

"But what about the special tools and jewellers' rouge and all that which will be required?" he said, "you being a blacksmith".

It so happened that we had a very old buffer wheel at work. It must have been bought for a special job years ago, and there it was covered in dust on a shelf and with it a piece of buffing cake (which was like a block of hard household soap, and you held it against the cloth face as the wheel spun round). The jugs were returned to the minister as good as new, and he was a very happy man.

Clifford Riley

Clifford at work

"Can you do anything with this?" said the lady who had just walked into the yard.

As she spoke, she was holding a battered old copper kettle in one hand, and a bent and broken stand in the other.

"It belonged to my husband's family and he is very fond of it, otherwise I would have thrown it away".

The handle of the kettle was porcelain with a thin spindle running through, fastened at each end with a small hexagon brass nut.

I could see that there was no money to be made on this job, but I wanted to help, so I said that I would see what I could do. On examination, the stand was not actually broken, but two soldered joints had come apart causing the whole thing to twist out of shape. It was a simple job to straighten it up and re-solder the joints, and it was as good as new. The kettle was going to be time consuming, so I took it home, where I had a little workshop in the underdrawing room. The porcelain was a problem, but suddenly I had a brain wave. The handles on the end of lavatory chains in the old days were made of pot. Would something like this do, I wondered? I made a tentative enquiry at a plumbers' merchant and - lo and behold - they had just what I wanted. Lady luck was on my side. I purchased 2 of these, and when I placed them end to end they made a perfect handle, and were just the right length to fit the brackets on the kettle. The bulges on the body needed a lot of patience and were quite awkward. The tool I found most useful was an old billiard ball. I could just get part of my hand into the hole for the lid, and with the ivory ball between my 1st and 2nd fingers I was able to hammer the bulges out gradually. The husband was over the moon with joy, his wife was pleased, but realised it was now another thing to get it cleaned. I was happy because they were happy, but this was a labour of love. You don't make a fortune out of jobs like that, but you do get a lot of satisfaction.

Wrought ironwork was really a sideline. I enjoyed doing it, but so much cheap stuff was being advertised in the Sunday papers of the time that most people did not want to pay our prices for individually designed and fitted gates and grilles. I had an extensive catalogue of designs and photos of gates and grilles that we had made, but one day I had to work to a customer's idea of wrought ironwork.

The local builder used to come to us for grilles to be fitted into doors of houses that they were building on a little private estate. We would make the grille to fit their doors and then their joiner would fit it into the door.

A rather fussy lady had had her own idea and we were only too pleased to oblige. Instead of a lot of fancy scrolls etc, she wanted it plain. She had in mind 3 metal bars, twisted along the whole length running from top to bottom. One in the centre and the other two equally spaced at each side. These were to be welded into a frame to fit the recess in the door. On completion, the joiner collected the grille and set off to fit it. An hour later, he was back with it.

> "Won't it fit?" I asked as he approached.

> "Aye, it fitted alright".

Then he burst out laughing, which was unusual for him.

> "I got it fitted," he said, "and stood back to look at it along with Mrs Smith, when a gust of wind came and slammed the door shut".

He laughed again.

> "You should have seen it. When the door banged the twisted rods vibrated like cello strings and knocked out the glass".

I couldn't help laughing as well.

Chapter 17

War work

Early in the war we received an enquiry from a firm of Boat Chandlers in Plymouth. Could we make some forgings for them? They had to be forged accurately, and subject to Admiralty inspection. We replied in the affirmative and said we would be pleased to quote on receipt of drawings. The package arrived by return of post, with the necessary drawings, and we saw that these articles were something that we could easily make. We sent in our quotation and waited. Imagine our surprise when, a week later, an order arrived for 75,000 to be supplied in batches of 200. The name and telephone number of the Admiralty inspector who would inspect each batch were also included, as was an Admiralty certificate for us to show in case we had trouble getting the material.

The title on the drawing was Dan Buoy Stave (metal work).

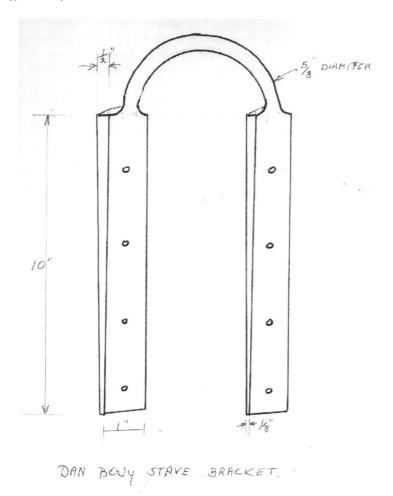

The Dan Buoy Stave Bracket

The job consisted of a U-shaped bracket, which was rounded in the middle and with the ends tapered down from ½" to ⅛" thickness. There were eight holes in each one to be drilled. There was another part which was a tube about 2" diameter inside, with a cap welded in one end. This piece had two holes in. The work was easy but monotonous, so piece work was agreed and we found out that when we had made the necessary tackle, one man could forge 100 in a day, once the flat bars were cut to length ready. We made a Jig, so that all the holes could be drilled in exactly the right place. We found that constant hammering of these narrow pieces (they were only 1" wide) made a groove in the anvil, and from time to time we had to build up the anvil face with hard surfacing rods and regrind the face.

When we had completed the 75,000, we received another order for 35,000 and that took us till the end of the war. The Admiralty inspector looked at each batch and then gave us his stamp to put on about 20, and then we dispatched them by rail to Plymouth. There, our customer (by the name of Lethbridge) fastened our forgings to wooden poles which they were supplying. We did not know what these were used for and during the war you didn't enquire too much about things like that.

Many, many years later, I attended a meeting of the Men's Fellowship at Church, when a former naval officer was telling of his experiences during the war in charge of a minesweeper. During the course of the talk, he described how they swept a path through the minefield and then threw the Dan Buoy overboard to mark a safe passage for the convoys. So, after all those years, I found out that all our good work was tossed into the sea!

It was a good thing to get onto the books of the Admiralty as suppliers of forgings, because one thing led to another. One day we received a request for leaflets and prices for our Riley chain repair link. We normally made them in sizes from ¼" to 1 ¼" diameter. The heavier ones were used in the stone quarries. Anyhow back came the reply, could we make these in the larger size of 1 ⅝" diameter? We had never made this size before, but said "yes we could" and from then onward we received regular orders for the 1 ⅝" diameter size for the repair of anchor chains. We had to make new tackle for these and they were too heavy to bend 'jump' the corner up by hand; one man had to use the 14 lb maul and act as striker.

What a pity that what was good enough for the Navy in extreme conditions was later to become illegal. Legislation was passed that broken chains could only be repaired by welding and our link was condemned.

It is strange how, when you are forced to try a new machine for a special job, you soon wonder how you managed without it. We had an enquiry from a firm in Sheffield who were making 'ratchet' handles. The handle was round, but the other end was square with a bulbous end. This end had a slot in it which had to be accurate, because the whole handle had to be machined, and the customer wanted to take off as little as possible in the machining. We had made similar things before, but not in batches of 50. We had previously cut a slot by hand with the Oxy Acetylene cutter and then shaped the slot on a mandrel. This method was no good in this instance, as we needed accuracy. We had seen an advert in the trade papers that BOC (British Oxygen Company) had a little machine that ran on a rail and cut accurately in a straight line, so we sent off for details.

The rep was soon here with a briefcase full of leaflets and prices of different machines. He was not very impressed with our choice of machine and suggested that a larger and more sophisticated one would be a much better buy. He suggested their 36" profile cutting machine. It had variable speed, would cut up to 6" thick and would not only cut straight lines but would follow a template with a magnetic roller and cut any shape you wanted. The rep was very persuasive and we placed an order for such a machine, although we only wanted to cut straight lines. What a boon that machine was and within a couple of months we wondered however we managed without it. It opened up a new avenue for us and we could give a fast service for profile cut parts like gear blanks, cams, flange plates, discs and anything that was asked of us. We produced gear blanks and cams for wire cutting machines, intricate shapes for card clothing machines, and flanges for the valve works. We eventually created a library of over 100 different shapes and sizes of pattern for all our different customers. We were able to build up that side of the business, stocking plates up to 3" thick, to such an extent that eventually we were forced to buy another similar machine.

Chapter 18

Home Guard Duty

I was 18 when war broke out, and it was not long before four of us from work made our way down to the local police station to enrol in the L.D.V. It was a little bit like Captain Mainwaring in Dad's Army. The police were unprepared for the response to Churchill's call to arms to defend the homeland. Enough volunteers from 18-60 enrolled and 5 platoons were formed. I was allotted to no. 5 platoon (railway section); Brighouse was a busy rail centre in those days and the railway sidings stretched right down to the end of Birds Royd, and filled the whole area between the high wall down Birds Royd and the bottom of the embankment at Healey Wood.

The Home Guards, with Clifford centre of back row

The goods warehouses were full of bales of wool for weaving into khaki, and goods were coming in and out every day. There were about 12 horse-drawn carts as well as the new vehicles which had one wheel at the front and 2 behind the cab. These were attached to various trailers and were called 'mechanical horses'.

The stables were along Cliff Road. All through the night goods trains were being made up and shunting engines were puffing away nudging loaded wagons along different tracks to make up the trains. Once the wagon had gained momentum, the engine stopped and reversed to sort out the next truckload. This was all happening in the dark and was very eerie as a truck glided silently down the track, and then crashed to a halt as it hit the buffers of the rest of the train being assembled. For safety reasons, non-railway employees like myself had to be paired with a railway worker for Guard duty. My partner was the signalman for the Anchor Pit signal box a mile down the line.

There were 8 of us on guard duty each night and we did 2 hours' patrol in twos from 8 o'clock at night to 6 o'clock in the morning. The 8 o'clock start gave those who were working late time to report for duty, and the 6 o'clock finish gave you time to walk home, change out of uniform and have a bit of breakfast before starting work. We were on guard duty every 8 days, so that we changed the day each week. Every Sunday morning we paraded for drill and instruction in the goods yard. We learned how to tell which line the main trains were approaching on, by noticing which way the wedges were driven into the shoes holding the rail to the sleeper. We also put our ear to the line. We had to know our way through the warehouses, and which route to take through the labyrinth of lines where trains were being made up. We were taught how to use detonators, which we could clip to the line in case of emergency. If the line should be damaged we were to proceed as far as reasonable up the line (in the right direction) and clip 3 detonators to the line at 6ft intervals. As the train passed over, these would explode and the driver would halt the train with all haste.

When we were on night duty we had a railway camping coach which had been adapted and placed in a siding. It was very cosy and had a little coal-burning stove in the centre. There was always plenty of coal available from the side of the plate-layers' cabin, where they had a stove in winter. The plate-layers were responsible for the upkeep of the track. During our 2 hour patrol we had to visit the passenger station, tour the warehouses, keep watch and be alert to any situation. We had to proceed as far as the Anchor Pit signal cabin, then from there back up the branch line to Clifton viaduct, and back to the cabin.

One night I was in charge of the patrol and decided I would do the last shift with my partner. At 10 o'clock the first patrol had not returned. This was most unusual, as they were always back on time. After about a quarter of an hour, I sent out the next pair to see what had happened, with instructions to report back. Midnight came and no sign of the 2nd patrol. Had parachutists landed? Had they been waylaid, even killed? It was a tricky situation. Not knowing what to expect my partner and I donned tin hats and fixed bayonets, loaded our rifles and set off down the line towards the Anchor Pit. As we passed under the bridge at the bottom of Woodhouse Lane, we noticed a glow in the sky. We got as far as the end of River Street, and there they were – all 4 of them – looking over the railings towards where the glow was coming from. Drury's soap works were on fire, and they had a grandstand view. I gave them a rollicking - and then joined them till the fire was eventually put out.

As things got organised better we started having a platoon night on Friday night for more instruction. There were various weapons being produced for the home forces, and eventually we were given denim uniforms. After about 6 months we were renamed the Home Guard and received proper uniforms with black gaiters. Boots came a few at a time, until there were just 2 of us without official boots. Mine were broad fitting size 12's, and a fellow who was a driver for Millers' oil and of very small stature required a size 2. Eventually a parcel of boots arrived at the shipping office in the goods yard (where the officer worked), labelled 'Size 12 Riley'. Much fun was made in the office because of the weight of them, and they were weighed and the weight posted on the notice board. There were expressions of sympathy at my having to carry these about, and commiseration to Laurence who still had not received his size 2's.

I did not open the parcel until I reached home after parade, and to my surprise the size 2 boots were fitted neatly into my size 12's.

One day the officer – Captain Mason – took me aside and informed me that he had made arrangements for me to spend 2 weekends in Harrogate, at a training camp run by the army. Here we were given weapon training and we learnt how to use the rifle, Sten gun, Blacker Bombard, Northover Projector, 36 and 68 hand-grenades, sticky bombs and Browning machine guns. I was promoted to Corporal and had to impart this knowledge to the rest of the platoon. The Sten gun was all metal with a magazine of 20 bullets. We had a practice range on Bradley Plains, where we were allowed 1 magazine each. You could set the gun to fire single shots or rapid fire. We were supposed to be able to fire 5 single shots with the setting at automatic. I was the only one who managed this exercise.

The Blacker Bombard fired a small bomb with a hole up the middle, and this fitted onto a spindle on the gun. The sight was very crude and consisted of 5 pins on a bar. Each pin represented about 10 miles per hour of the tank you were aiming at, and if you could estimate his speed you had a hit. The Northover Projector was a tube like a small fall pipe. You put the charge in from the mouth, and then inserted either a phosphorous bomb (liquid in a bottle) or a 36 hand grenade. The sticky bomb was the size of the ball cock in the toilet tank. It was a spherical shape covered in a substance like treacle, and had a short handle. This could be thrown at a tank, and it would deflate like a balloon and stick to the tank. The 'Browning' was the proper thing, mounted on a tripod and with bullets on a long belt and fed through the gun by another man.

As we passed out in these various weapons, we were issued a little red patch to put on our sleeve.

When I was younger I often watched the Territorial Army shooting on the rifle range at Toothill; from New Dick you had a grandstand view. I never expected to be shooting there myself, but the occasion arose just once. We were each allocated 10 rounds and were supervised by the army. I had never used a rifle before but followed instructions and did well. I got 9 bulls eyes and 1 inner all in a 4 inch circle. The officer tossed me half a crown, and said

"Good shooting young man, keep it up".

As the war progressed it was decided to put a battery of anti-aircraft guns at Southowram. A large field to the right of the Packhorse Pub was commandeered. 64 rocket projectors were installed. These covered a large area and were to be fired all at once; making an acre of death in the sky should a plane venture into range. It was not long before these were deemed obsolete and were replaced by twin projectors which could fire 128 rockets into the air at the same time. They were never fired in anger. These were manned by the Southowram Home Guard, but others were recruited from the Brighouse Battalion to train and be reserves if needed. I was one, and did a couple of weekends' training at Southowram.

I was then promoted to Sergeant, and had my own squad.

It was not long after my promotion that an exercise was arranged. My squad was to attack the station and goods yard and the other 2 squads were to defend. Army officers were to observe. This was to start at 10 am on a Sunday morning. I had the use of Mr Copperwheat's coal wagon to transport my squad to attack from about a mile away. We disembarked at Bradley Bar, blacked our hands and faces and set off through the fields. It was my idea to get to the embankment and then open fire down onto the defenders from the top. We took our time, went very carefully, crawled behind hedges and arrived at our designated point. To our dismay the rest of the squad were dug in waiting for us, and it was deemed that the attack had failed. At the inquest later, the officer was full of praise and said that I had done well to get my squad in position without anyone being seen. The reason we were caught was that one of the scouts had noticed Mr Crabtree's cows all walking in line down the edge of every field. The defenders never saw a movement or a sign of us other than the blessed cows. It was congratulations all round, and deemed a very good exercise with an unusual twist.

As the war progressed, the patrols down the line were discontinued and we did our night duty at the drill hall down Wakefield Road. This was well received as we had camp beds and could snatch a few hours' sleep.

In the years when our Country
was in mortal danger

CLIFFORD RILEY

who served 8 July 1940 - 31 December 1944

gave generously of his time and
powers to make himself ready
for her defence by force of arms
and with his life if need be.

George R.I

THE HOME GUARD

Home Guard Certificate

Chapter 19

A Woolworth's Hacksaw Blade

Before the war there was a little shop in Huddersfield Road, opposite Whiteley's Paper shop (now a travel agents), which sold Ariel motorcycles. Often, after leaving Park Sunday School (now Weatherspoons pub) and waiting for the tram home to Rastrick, clutching my halfpenny fare in my hand, I would gaze at the chromium-plated petrol tanks and shining paintwork in the window. It wasn't that I was hoping to have one as I got older, no, it was the fact that they appeared to me such things of beauty. Years earlier my uncle had taken me for a ride in the sidecar of his combination and that must have sown the seed in my mind!

A Village Blacksmith

Frank Holdsworth and Clifford

It was 1945 when I applied for and received my first driving licence. It was pink in colour and titled Provisional Licence. It entitled me to drive anything from a motorcycle to a road roller; I was now ready for my first motorcycle. My two mates, Bob and Ken, were also contemplating owning a bike of some sort. I heard that Derek Spur had one for sale, so off I went to have a look. Derek's father owned the Bull Fold Garage and in his spare time Derek could be heard playing his violin in the inner sanctum; a good player he was too. The bike he was selling was a Triumph Tiger 80 with chrome tank and silver mudguards. I knew I had to have it even though I had never been on a motorcycle before.

> "Nothing to worry about", said Derek. "Hop on the pillion seat and I will show you how to change gear as we go along."

We rode to the end of Churchfields Drive, up Bonegate Road and Garden Road to the top of the Recreation Ground, back along Granny Hall Lane and back to Churchfields. Not much more than a mile and the duration of a few minutes.

> "You will be OK now", said Derek and wished me well as he counted the money.

I was on my own and did I know it! I set off in bottom gear and got as far as the top of Bridge End, Rastrick when, horror of horrors, the thing spluttered to a halt and I finished up pushing it home.

During that week my parents had been on holiday but were due back the following morning. They were not aware of the purchase and then, because it had conked out, I was berated for buying a 'pig in a poke'[19], as father described my activities! Derek was just setting off on holiday himself when I phoned him for some advice. He called in at the Smithy,

---

[19] A useless object

put his foot on the starter pedal a few times and diagnosed a serious problem.

"You'll have to strip it down and see what's happened", he said.

Then off he went on holiday!

I did as he said and found the trouble very quickly; the rocker arm had broken. Luckily I was able to weld it together, reassemble the parts and it started first time. It turned out to be a good experience and the bike never gave me any trouble again.

It was not long before there was a shortage of petrol and one had to conserve it to keep on the road. By this time all three of us lads were proud owners of motorbikes and we used to set off at every opportunity. Ken got a Royal Enfield and Bob a Francis Barnett. The latter was a small two stroke engine and went pop, pop, pop, achieving a top speed of 30 miles per hour flat out. We called it the 'frantic Barnett!' We were all indebted to our friend Mary, whose father owned the local taxi business, as they had a petrol pump of their own and special supplies for the hearses. Mary could usually spare a gallon to top up our tanks. In the end, though, we were not allowed on the road and the bikes were laid up for three months.

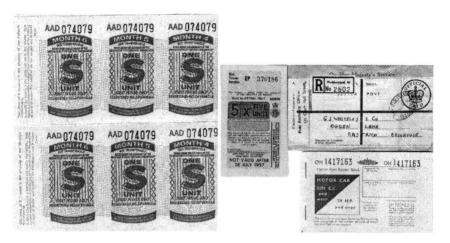

Petrol rationing.

Once petrol was available again we were back on the road and in 1946 had an idea that it would be an adventure to spend a holiday abroad and, for no special reason, Switzerland came to mind. Just about this time Triumph came up with a completely new model, a 350cc twin cylinder engine, which purred like a cat, was quiet as a mouse and went like the wind. It was so easy to start that you could press the kick starter by hand and it would go. The frantic Barnett was no good for Switzerland anyway, so Bob and I set off for Leeds and ordered a couple of new 350 twins, our old bikes being taken in part exchange. We were now ready for anywhere.

We started planning our trip and, eventually, in 1947 the three of us and Bob's cousin David set off for Switzerland. My mother was interested to know which way to go to Switzerland. This surprised me as she had never been abroad in her life. I said,

> "Well we get the ferry at Dover, then across to Calais and then…"
> She stopped me there.

"I meant which way out of Brighouse will you go?"

"Down Wakefield Road to Dewsbury, then Doncaster", I answered.

"Well fancy that", she said. "Whoever thought you could get to Switzerland going down Wakefield Road!"

We had done a lot of preparation before setting out. I had made carriers for us all for stowing the luggage and fitted extra tool boxes for spare parts. Before I finished work on the Friday, Harry, an old employee, who was still working at 66, gave me a small 6" hacksaw, which he had purchased at Woolworth's some years before and had no use for any more.

"It might come in handy", he said. "You never know what might happen on a trip like this."

How right he was! And what a treasure that 6d [20] hacksaw turned out to be, with its one blade and no spare.

The local holiday week started on the first Saturday after the second Thursday in August and off we went aiming to get to Canterbury that day and then do the short trip to Dover in the morning. About ten miles from Canterbury my legs started feeling very warm and I realized the engine was running hot and getting slower. The other lads disappeared from view and my new Triumph seized up and came to an abrupt halt. Eventually they noticed my absence and Bob appeared on the horizon

---

[20] In those days everything at Woolworth's stores cost either 3d or 6d.

looking for me. We felt there was little we could do at the busy roadside, so we attached a tow rope and Bob towed me to the hotel in Canterbury, where we had arranged to stay the night.

This image is the scene of the repair on the road to Arras

During the evening we pondered what to do and were still in a state of indecision in the morning, when it was time to set off. My engine had obviously cooled down overnight and we managed to make a start down the Dover road. It was not long before my worst fears materialized, however, and once again the engine overheated and came to a shuddering stop. The time for embarkation was drawing nearer, so I was hooked up behind Bob and towed to the port. We felt we had to go and would try and sort the problem at the other side of the Channel.

Of course there were no roll on, roll off ferries in those days and everything was loaded over the deck. A sling was wrapped under the handlebars of the bikes and they rose precariously in the air to be stowed in the hold. Cars did have a shoe under each wheel and made a much more dignified passage!

Loading the Ferry, Ken, Bob and Me

The journey was not at all enjoyable, wondering what awaited us in France, but we assumed that if not too much damage had been done to the engine it would at least start again and get us somewhere where help was available.. Fortunately it did start and we managed a few miles from the port on the road to Arras. It was not too busy and there were grass verges on each side where we came to a halt. There was no garage or any sight of habitation of any sort, so we had a confab [21] and decided that there was no option but to dismantle my engine and, if we could find the fault, to try and mend it.

It was to be a major operation and we prepared ourselves for the worst. Off came the cylinder head and we could see that the top of one piston was rough and discoloured, so we knew where the trouble was. The only option was to remove the engine block and leave the pistons free. This we did and found that the metal rings on this piston were embedded in the aluminium piston and it looked a mess.

---

[21] Discussion

"Not to worry", I said. "I have a spare piston, so all we need to do is replace it and we are on our way again."

Not so! We could not remove the piston, because it had fastened itself to the gudgeon pin, which went through the connecting rod, and we could not shift it. After much discussion I had a 'brainwave', which did not rouse much enthusiasm from the other three, but as they had nothing better to offer we went ahead.

My idea was to saw the piston into segments, like an orange, and remove it piece by piece. I produced my Woolworth's hacksaw from the toolkit.

"You are on your own with that!" said the others. No-one dared run the risk of using the saw and breaking the one blade we had with us.

We stuffed cotton rags around the piston to stop bits dropping onto the sump and I set to work. It was coming dark by this time, so the lads put their bikes in an arc round me and switched on the lights. Patience paid off and eventually the piston was removed from the connecting rod and the gudgeon pin. The pin, however, was still fast in the brass bush in the top of the connecting rod. We managed to hammer the two apart and it was here that we found the rest of the trouble. There was a hole in the top of the connecting rod to feed oil to the gudgeon pin. This had to pass through a hole in the brass bush, which should be a tight fit in the connecting rod. It had worked loose and twisted round, thus cutting of the supply of oil. We took it in turn, rubbing the brass bush on emery paper until it would turn freely in the piston; then with the hacksaw we made a series of spiral grooves so that the oil would ooze through, whatever the position of the bush. All that remained to be done was to put it all back together again and hope the engine would start. It did! We arrived at the hotel we had booked for the night, in time for breakfast!

The Triumph lived up to its name after that, but I did treasure the 6 penny Woolworth's hacksaw.

In Switzerland

Not long after the Swiss trip a local bike enthusiast came round and made me a proposition. He was doing a lot of miles and, petrol being scarce, he offered to exchange his powerful Norton for my Triumph. I told him of its condition, but he was not deterred. He said my Triumph plus £30 against his Norton and we had a deal. It was too good to be true and we shook hands on the deal.

What a bike that was; 500cc engine and telescopic front forks and it went like a bomb! It was surprising how often it was needed at work; one had to wonder how we did all the errands before! Fetch this! Fetch that! Go there… it was so convenient. We were selling quite a few hydraulic trucks in Lancashire at the time and sometimes one would fail and I had to go round and repair it. I would set off all clobbered up with my tools on the back and arrive like somebody from outer space. I always made the excuse that the van was in for service!

The usual trouble with the trucks was that they topped the oil tank up with old, dirty engine oil; muck would get under the valves, which were steel ball bearings and which rested on a hole in the pump, causing them to leak. I had to strip the pumps down, swill them out, reseat any leaking valves and fill with clean oil. I felt it would have been a lot better to have arrived smartly dressed in a smart little van and changed into clean overalls on the job, but that was wishful thinking.

Chapter 20

Nora

In 1950 I realized that there was more to life than work and motor cycles and got married to Nora. I was 29 years old at the time, yet my stepmother wanted to know what was the hurry? Marriage was a wonderful experience for me as I was introduced to a loving family and, contrary to popular belief, my mother-in-law was a treasure.

Nora

Nora was one of five sisters and two of them were to marry two of my pals, Eric and Bob, so we became a close family of brothers and sisters.

Clifford Riley

Later, when I took over the business, Nora gave up her job of school secretary and became office manager. She got all the books ship shape, bought an adding machine (which she treasured), passed her driving test first time and made small deliveries, answered the phone, paid the wages and became my indispensable helper.

Like me, Nora had musical parents. Her father was a teacher of the violin and viola and her mother played the violin. Soon after we were married we were encouraged to play an instrument and Nora was designated to play the cello. She had lessons from a family friend and did very well. I sat in on the lessons and decided to try myself! I did very well too and my father then presented me with his cello, which he was now too old to play. Although I didn't play for a long period of time between 1963 and 2004, I was able to start again and now have many hours of pleasure from my music.

Still playing the cello

Chapter 21

From Bikes to Vans

Soon after Nora and I were married I got tonsillitis very badly and had to have my tonsils removed. I was advised to part with my beloved motor bike. Fortunately I had no difficulty selling, as my brother-in-law Ray could not wait to buy it, so it was around for a long time.

Having just got married we did not have a lot of money to spend and I did not think we could afford to run a car. My wage was £6-14-0 a week and Nora had given up her job when we got married; that is what most women did in those days. My parents were horrified when, sometime later, Nora took on a part time job as secretary at Longroyd School. However, with the money from the sale of the bike and withdrawal of a few savings, we bought a Raleigh three-wheel car, with the single wheel at the front, for £80. The starting handle went in at the side of the bonnet and the engine was from the Austin 7. It had a folding hood and during the summer it was nearly like being on a bike, but with the protection of a windscreen. We had this for quite a while, and then some fool told Nora that they had a habit of turning over and she was unsettled after that. We had to put it up for sale. Nora cleaned and polished it and we sold it to the first person to enquire for £90; the only car we ever sold for a profit!

We bought a 1933 model Morris Minor for £100. It belonged to a man from the Gasworks and it had always been owned by someone in

Brighouse. The engine was well worn and there was always a smell of oil. Nora was never a good traveller and told me she could not put up with it, so I took it to the local Morris garage and naively asked if they could cure the smell. I never thought to ask the price; it seemed trivial really. They cured it alright! They rebored the engine and fitted new pistons at a cost of £47. That came as a shock!

Our next venture into the car market came in the shape of a Jowett Saloon on the market for £120. Nora now had a part-time job so we had a bit more money to invest in a car. It was very roomy for a 7 horse power car and, being made in Bradford, spares were easily available. The cushion on the back seat could be taken out and this made it easier to carry things from work. I was to regret this!

One time, when my father had posted out a handful of leaflets about our Elevator Truck, we received an enquiry from a firm in Bradford. Would someone go and give them a demonstration of our Beam Truck? This was, of course, intended to be used inside a woollen or cotton mill on a flat floor. I noticed that the firm dealt in carpets. I was stricken with foreboding; it was not designed to deal with carpets.

> "Well let them look anyhow", said Father, as I endeavoured to get the machine into the back seat of the Jowett. It weighed about 250lbs and proved most difficult even with help.

On arrival at the mill I made myself known and the manager accompanied me to where I had parked the car. I foolishly expressed my doubts about it being of any use to him, but he insisted I demonstrate and pointed where I should go. I unloaded and pushed the truck to the appointed place, which was cobbled and where a wagon was just delivering some carpet rolls. I drew alongside with the truck and elevated it to the required level for a roll of carpet to be rolled onto it. I then lowered it to the bottom position and tried to push it to the loading bay. The manager was most arrogant and told me in no uncertain terms how I was wasting his time and what he thought of the truck. Of course, the truck would not budge on the cobbles, much to the amusement of the assembled staff, who had to lift the carpet off and carry it away. Then they all watched as I struggled to get the truck back to the car park and into the car single-handed. All in all a very humiliating experience, arriving in a clapped-out old car and being made to look a fool in the bargain. I vowed never again.

I think my father saw the point and decided (with a little bit of prompting) that we should have one of the new Hillman Huskys[22] and run it on the business. This took a load off my shoulders. The Husky had a 1000cc engine and a back door that opened onto a reasonable space when the back seat was folded down. It would easily carry one of our trucks and it gave us good service.

A few years later a new and more powerful Husky replaced the first one and we were able to update at very reasonable cost. The exchange was due to take place on a Monday. Over the weekend we had been visiting an old Aunt and it came up in the conversation that we were changing the car the following week. We explained that it would be a similar car, but green, whereas this one was grey. Her face went ashen,

> "Don't you know that green cars are unlucky", she said. "You should never have a green car".

---

[22] a new type small shooting brake

We took delivery of the new car on the Monday, as arranged, and we were very pleased with it. As I drove it into the garage later that evening, the wind caught the garage door and slammed it against the side of the car. It caught on the door handle, broke it off and dented the door at the same time. That was just the start.

Some time later it was parked in Ogden Lane outside the works, when a taxi driver attempted to do an emergency stop and hit the accelerator instead of the brake. He ran into the back of the Husky and knocked it onto the pavement.

Twice a week we had deliveries of oxygen and acetylene cylinders. For every one we received we had to send one back. There were no hydraulic lifts in those days and loading the cylinders took three men. The cylinder was wheeled, on a sack cart, from the welding shed at the bottom of the yard to the roadside in Ogden lane. The cart was tipped onto the front lip and the cylinder removed in the vertical position. The driver took hold of the neck and two other men, with a bar of metal, with a half circle pressed in the middle to the shape of a cylinder, put this under the bottom end, and the cylinder was lifted horizontally. At the given word of the driver, the load was swung and lifted and the cylinder guided onto the wagon. This process had been going on for years and everyone knew exactly what to do.

The smithy yard showing garaged cars, c.1990

On one particular occasion the Husky was parked down the yard while loading was going on. While the cylinder was being held in the horizontal position, ready for swinging onto the wagon, the sack cart began to move, backwards way, down the slope. The three workmen stood there helpless, holding the cylinder, as the cart gathered speed down the incline and ran straight into the back of the Husky, making two dints in the rear door and twisting it out of shape.

One very cold winter's day, when the ground was covered with ice, I was parked on Back Park Street, behind the Post Office. When I was ready to go I found it impossible to turn round because of other parked cars, so I decided to go to the end of the street, turn left under the arch, and into Commercial Street by the butcher's shop. I skidded on the ice, misjudged the corner and scraped the passenger door on the corner of the wall.

Well, I had been warned that green cars were unlucky! Never underestimate an old Aunt's tale.

Since the cars were no longer mine, but belonged really to my father (although he couldn't drive), Nora and I felt obliged to give my parents as much pleasure from them as possible. To this end we took them for a run most Saturday afternoons, when we would have a picnic or have afternoon tea in a café somewhere. As they got older they had difficulty getting in and out of the Huskys, so we made a big decision; we would get a larger car for pleasure and a small van for work. I went down to Bates' garage on Bradford Road and negotiated the purchase of our first Morris Minor van.

On my first drive I felt boxed in and a bit claustrophobic, but it is something one gets used to. Like everything else, when you get a new toy you wonder how on earth you managed without it. One of the selling points of the van was that it had a heater. It was about 10" in diameter and fitted under the dashboard on the passenger side. How I appreciated that. It did good service, but with a capacity of only 5cwts we thought that our next one should be a bit bigger. Our van never did big mileage and never really earned its keep, but yet we could not do without it. It turned to our advantage in the 1950s when petrol rationing came in again. We got the same ration as a van that was on the road all the time, so we were never off the road and had coupons to spare.

We next chose a sturdy Morris Oxford van with the gear change on the steering column. That was new and took a while to get used to, but then I came to really like it. We did a lot of work for Hattersley's, the valve works on Bradford Road. Valve flanges were cut from plates and we supplied them by the thousand. Sometimes castings did not always turn out perfect and often we were called upon to weld these. New machines were coming into use that were controlled by a magnetic tape and these completed work in a fraction of the time that a man took to do the same job. There was one drawback; sometimes there was an error in setting up and before it was noticed 20 valves may have been wrongly machined and we would have to weld them, or build them up, to be re-machined.

It was on such an occasion that I had a brush with the law. Twenty valve faces had been grooved too deep and had to be built up with welding. When the job was complete I loaded them into the van and out of curiosity I weighed one. It weighed just 56lbs (½ cwt); that meant that my van was loaded to capacity exactly. As I drew into Hattersley's yard a car followed me in. As I got out of the van a well dressed man and woman approached me. The man showed me his identity card and said,

> "Customs and Excise, we have reason to believe that you are overloaded. You will follow me to the weighbridge on the trading estate and my assistant will accompany you."

I told him he was making a mistake, but he was not interested.

On reaching the weighbridge I parked and got out of the van.

> "Get back inside", he ordered and weighed the van with me in it.

That was a surprise to me as I thought a 10cwt van could carry a load of 10cwts, but that was not the case. He told me by how much weight

I was overloaded and it was 18stones, just my weight. He took all my particulars and then asked who my boss was. I said I was.

> "That means as a driver you drove a van overloaded and being the Boss you allowed it to be done! Guilty on two counts. You will hear from us", he said, and I was allowed to proceed.

I did hear from them. They took into consideration that I did not drive dangerously and that, in ignorance of the law, I had taken the trouble to check the weight of casting, which was true. Also taking into account that my weight was the amount overloaded, they had decided that, under the circumstances, they would let me off with a caution. Once again it seemed that the van was not big enough for our requirements. This always seemed to happen. In the end we had to purchase a 1 ton capacity van. Sad to say, even with this, we had to take some risks, because of the urgency of the job, and we were guilty of overloading from time to time.

Chapter 22

My Father

I was groomed from an early age to be a blacksmith, although in later years my father used to say that I could have done anything I chose. I never felt that that was an option. He was a typical Victorian father. I can remember the resentment of having to be in bed by 9 o'clock and then lying awake for an hour or so hearing other youths enjoying themselves. Later, I used to be one of the first to leave the Tennis Club and pedal home like mad to make the 9.30 deadline. The war put a stop to that. Late nights on Home Guard patrol and attendance at Technical College were commonplace, as was being out all night on duty. One day he gave me a wigging for talking to girls in the church doorway after the service, even though he knew them all and most of their families. That was the last straw, and as I later donned my uniform to go on night duty, I turned on him with a vengeance. I said that I was going on night duty now, that I would have a pint with the rest of the lads, then between patrols I would join the card school and gamble the night away. I looked him in the eyes and said,

"And you cannot do anything about it".

He went white, but I took full advantage after that, enjoying late nights and early mornings when the opportunity arose. I could patrol the railway with a sub-machine over my shoulder, learn bayonet fighting and all father was worried about was girls in the church door.

We did not have a good relationship; the only topic we had in common was work, and we often disagreed about that. I sometimes felt that old Mr Brown had a point when he said that my first day working for my father was the worst thing I would do. The first wrought-iron work that I had to do was for a pub sign. I had made scrolls on the end of spout hangars, but these being at roof height did not get close scrutiny. I was having a bit of a struggle setting it out, so asked father for advice. He took a look at it, and then said to me (words that I remember well to this day):

> "You must try to work it out; you will not always have me to tell you how to do it".

He lived to regret those words, as I never asked his advice again as long as he lived. I'm not proud of this, and as I look back it is with a feeling of regret that it happened, but you cannot turn the clock back.

After the war, father decided that he would make me a partner in the business and that I could have a ⅓ share. It made not a scrap of difference, and I still received the same wage as the men, although by now I was going back at night and doing the wages and any quotations that required attention. All the book-keeping was done in the family home on Stackgarth, so it was going home again for me to do those chores. If I took too great an interest in the ledgers etc, my father would craftily close them and shift papers over them so that I couldn't see. Later, as he started muddling, I suggested that he had only to say the word and Nora and I would take over the book-keeping, but he went on and on. Eventually, metrication became a must and he decided we should take over. We sacrificed a week of our holiday, and turned a store room into an office. On the Friday we were ready to take over, and went to the house to collect the books. My stepmother was typing a letter. There was silence for a while, and then I ventured:

"We've come to collect the books".

"Oh. It's alright," he said, "We've decided to carry on".

"Oh no you haven't," I said, "we've just sacrificed our holiday, and got an office ready at work",

And I proceeded to collect the books and did several trips to and from the works. Lastly, I had to take the typewriter from under mother's fingers. I asked where the petty cash was. Father put his hand in his pocket, and pulled out a 10/- note which he put in my hand.

"That's it".

When we eventually sorted things out, there was not enough money in the bank to pay the wages, and Nora and I had no option but to pay the first month's wages out of our own savings. Our first job was to round up all the late payers, and insist on payment at the end of the month.

Father still kept coming to work, however, to see what was going on and do odd jobs. We made a point of leaving the office tidy when we had finished work, and there was not a book or a paper in sight. When he felt he could no longer come to work, he died two weeks later.

During his life he had acquired the cottages adjacent to the shop in Ogden Lane, and also on one side of the road on Stackgarth. The average rent they were bringing in was about 3 or 4 shillings a month. He had kept no strict records. The rent money would go into his pocket and any expenses would be paid with a cheque on the works. He said that it didn't matter anyway as they were all his, the works and the houses, so there was no need to keep it separate.

The backs of the houses on Stackgarth from the Grammar School

The taxman had a field day. There was no record that I was a partner in the business at all, and no dates either. He came to the conclusion that my father was a clever evader of taxes and did an in-depth investigation. It was a few years later before everything was legally sorted out. In the meantime, the tenants in the two cottages in Ogden Lane demanded alterations and it cost £800 each to have a toilet and shower fitted. Further trouble was on the horizon. Brighouse Town Council decided to adopt various streets which were previously un-adopted. This meant that they would resurface the street and the pavements and charge the cost to those with a frontage onto the street. Panic stations. I arranged to see the solicitor, Mr Hinchcliffe, who was guiding me through the consequences of my father's death, and he (being an alderman and councillor) was aware of the Council's intentions. He looked grave. Apart from 1 house, I had inherited the whole of one side of Stackgarth.

Houses on Stackgarth from the smithy yard

"You must divest yourself of these properties, or you will be bankrupt without a doubt, and you must make the business into a limited company".

The houses now being in my name, I set about selling them. They all had sitting tenants. I sold the two in Ogden Lane for £1000 each, so there was not much profit there. It was a start, anyhow. The old family house was No. 4 Stackgarth; the next door was No. 6. Owing to the fall in the land there was a cottage underneath 4 and 6, this was number 2. The bedroom was under no. 4 and the living room under no. 6. It was occupied by an old gentleman who was quite happy there. One day social services came to see him, and said he could not live in conditions like this as one wall was damp. They condemned the cottage and allocated the old man a flat in Field Lane estate, and he had to get rid of his dog. He was very upset about the whole thing, and before he was moved he died.

I was now left with 2 houses sitting on top of a condemned one. Another thing, there was no boundary fence between the back gardens of 2-4-6 and the rear of the works. It hadn't mattered before, but it mattered now, so one had to be erected. I had a word with a builder who was a good customer of ours, and mentioned my problem of the 2 houses sitting on a condemned one. He came up with a good idea: knock a doorway through the cellars of 4 and 6 into the 2 rooms of no. 2. Then make up the door between the 2 rooms of no. 2 and make an exit door in the bedroom of no. 2. Thus, you would lose no. 2 as a house, and no. 4 and 6 would have an extra cellar and a door into the garden.

"Could you do this?" I asked,

And the answer was, "Yes, of course".

The job was done within a week and in 2 days the through-draught of air had cured the damp. The builder's bill for this work was £120.00. That's the end of that, I thought. But no! I received a letter from the Council wanting to know why I had proceeded with this work without planning permission. Because of the stupid position they had inflicted on me by condemning no. 2! They later sent me a form to fill in, and did give permission. I was able to sell 4 and 6 for £1000 each.

Not owning no. 8, there was now the question of no. 10 Stackgarth. This had access to the works property, and had to be fenced off. This I offered to the tenant, a lady who popped in occasionally to clean the office. I asked £1000, which she accepted – but with just one snag. The Council would only give her a £900 mortgage; could I lend her the remaining £100? I am still waiting for it to be repaid.

That just left a few yards on Stackgarth which abutted the property belonging to work. Fortunately, Calderdale came into being and Stackgarth has never been made up. In those days, it was published in the paper how much people had left in their will. When my father's assets were all added together, his house, the works and all the stock, it seemed a considerable sum to anyone not knowing the circumstances, and just reading the amount in the local Echo. I later heard that during discussions at the local bowling club, I was referred to as the Gentleman Blacksmith.

After my father's death, we built a new office at the bottom of the yard. That's when Nora gave up her job as school secretary at Longroyd School and took charge of all the office work.

Chapter 23

Ministry of Supply

On one or two occasions we had to put jobs out which required a press rather than a hammer and one day I was asked by the Ministry of Supply to quote for 75,000 Hot Pressings of similar shape, but in different sizes. They were to be supplied in large numbers from a very special steel. The steel would be supplied to us free of charge; we had to do the forging. I thought long and hard about this as, if I got the estimate right we could make some money, but if I got it wrong we could be working for nothing and might go bust.

There was a press on the market, which was just what we needed, so I decided to take the plunge and buy it. It exerted a pressure of 50 tons per square inch and could be used for hot or cold work. There were a lot of regulations about presswork and although the makers said this one needed no guard because of its slow movement, I wanted to be sure.

Because of our previous experiences with the factory inspector I thought I would contact him first, rather than wait for him to catch me out. He was soon on the job. Of course the press needed a very sophisticated guard – one that opened and closed completely as the foot pedal was operated. I said that if we did that, I could not use it for the purpose for which it had been bought. I explained how we were quoting for a job where hot pieces of metal were held in a pair of tongs, and that they would need to be done quickly as we would heat 6 of these at a time.

There would be no time to mess about or else the last one would be cold before we could get to it. Very reluctantly, he agreed that if we set the machine so that the gap between the press tool and the work was no more than 1/8", and we did only hot forgings (no cold pressings), then we could use the machine as supplied. This is where health and safety has gone mad.

The quotation for 75,000 pressings was accepted. I took a risk because I had no previous experience in pressing or in making press tools. These had to be made of very special steel to withstand the heat and the wear and tear. The job itself was very simple. There were 4 different sizes but of the same U-shape. The smallest was from 2" wide by 1/4" thick bar; and the largest from 3" wide by 1/2" thick bar. The bar was made of special material and very hard and springy, and would be supplied to us cut to the correct length which we had to specify after submitting various samples made from ordinary material. We were given a certificate to say that G.S. Whiteley (Brighouse) Ltd was an official supplier to the Ministry of Supply.

I was a bit apprehensive about the job because a few pence under or over the cost of forging could either make or break us. First of all, we had to make 4 press tools. That cost about £2,500 before we started. The press had cost £1,750. We started off heating the first batch in the forge. We had to have such a big fire that we set the roof timbers on fire near the chimney. The men were dripping with sweat and exhausted at the end of the day, and we were not making the forgings fast enough to make a profit. I contacted the Gas Board. Two men came over from Huddersfield and discussed our problem. They were very helpful and said they would go back and design a gas furnace to suit our purpose, and if we agreed a price they would make it for us. This was agreed upon. We made some trays out of stainless steel, and these would hold 4 or 6 depending on the thickness used. We welded a bracket on the front so that we could lift them out with a 5 ft long metal handle. It needed to be so long because of the heat from the furnace when the door was opened. The door was lifted by a chain over a pulley with a counterweight to lighten the load. The gas was forced in with a big fan and burned £3 worth of gas per hour. We could put 6 trays of plates in at the same time.

At last we were in production. The plates were supplied to us at first at about 200 at a time, but later they came in batches of 500 in old oil drums. We could not handle these properly, so there was only one answer. We realised we should have to have a forklift truck. We talked this over with the supplier, and decided on a small one where the man walked behind and manipulated it from a set of handlebars. All the lads were keen to see the new truck in action and have a go. Horror of horrors, I manoeuvred the truck into place, pressed the lever to 'raise', and waited. The plates stayed where they were and the truck sank into the yard. Our yard was composed of compressed clinker from the fires, it was no motorway surface. We had to buy a bigger, more expensive model with larger wheels.

Once again we were in production – for a time. The new truck was great, but it churned up the yard and made it a quagmire, then got bogged down. Would we ever master this job? There was no option – we had to have the yard concreted. The expenses were going through the roof!

On occasions when it was inconvenient, but urgent, our suppliers asked us to collect the plates in our van. It was only a half ton van, was obviously overloaded, and had a struggle to get up Lower Edge hill (we had to deliver to Elland). The suppliers found this method of delivery a big help, and therefore decided that in future they would give us a telephone call and we could always collect and deliver. We had to get a bigger van!

At last everything was under control. We were turning them out in their hundreds and finally making good money.

It so happened one day that the director of the firm in Elland visited our works about another job. His firm were involved in the machining of the work. They received the original bar from the Ministry, and then sawed it up and we collected the pieces and did the pressing. We then delivered them back and they machined them all over. We had to invoice them for our work. He paused to see how we were making them and was impressed by the speed at which we were now making them. His comment was,

> "You must be making a fortune out of me, the price you charge".

Sometimes you wonder – is it all worthwhile?

Chapter 24

The Riley Link

The Riley link was an open link, which could be inserted into a broken chain and then closed up cold. It had a punched eye at one end and the other end was swaged down to fit the eye. When the link was closed, the swaged end went through the eye and was riveted over, making a permanent repair, which followed neatly the contours of the chain. It was made in different sizes to fit all chains: from ¼" diameter to 1 ⅝" diameter. During the war the large size was supplied to the Admiralty for the repair of anchor chains, and any other chains that might need a repair.

The Riley Link

We ought to have made a fortune out of this, but my father never charged enough and in his later years I found that he was charging less than the cost of making them.

Production started in a small way, making a few here and there for the local quarries. One day a magazine dropped through the letterbox. It was entitled 'The Timber Trades Journal'. Why anyone would think it would interest a blacksmith is hard to tell, but my father put it aside, and in an idle moment he picked it up and read it through. He then had a brainwave. There were many adverts for logs and logging equipment in the journal, and Father pictured loggers dragging logs through the forest miles from the nearest blacksmith. Often, logs got trapped and chains broke. Perhaps our link would be of use to them. He took a gamble and had some leaflets printed, and then started to circulate them to the timber merchants whose names appeared in the adverts. Gradually the orders came in and sales really took off – eventually we were making 10,000 a year and supplying repair links the length and breadth of the country.

There was an offshoot to this, which was quite unexpected, as firms enquired if we could supply sling chains and snigging chains made up to their particular requirements (snigging chains were for pulling logs through the forest until they could be loaded onto wagons). This was before hydraulic grabs came onto the market. Of course we could, and this opened up another door.

We used to buy 100 ft lengths of chain from the chain makers in Sheffield and store it in 10 gallon oil drums with the top cut out. Soon word spread that we stocked chain, and quarries and other users came to us for their chains. Soon we had a stock of chain from ⅛" 'light jack' chain, to ½" diameter for heavier work. For this work we bought drop forged rings and hooks from the Sheffield forgers, and soon built up a stock of these.

We fitted the hooks and rings with welded links rather than our repair link. The making of the repair links was quite monotonous so there had to be some incentive for the men. We decided on a piecework scheme so if the men 'cracked in' they could make 133% of the normal rate. Also, we only made them in the morning, and then changed to other jobs in the afternoon. Prices were worked out in dozens, and men worked on them from 7.30am to 12pm. Different sizes had a different price. It transpired that in the 4½ hours the rate of production was:

- ¼" diameter = 5 dozen per morning (these were fiddly)
- 5/16" diameter = 6 dozen per morning
- ⅜" diameter = 7 dozen per morning
- ½" diameter = 6 dozen per morning
- ⅝" diameter = 5 dozen per morning
- ¾" diameter = 4 dozen per morning
- ⅞" diameter = 3 dozen per morning
- 1" diameter = 2 dozen per morning

Sizes between 1" and 1 ⅝" were not in great demand and were made at day rates.

There was a specific cutting length for each size and this was important. Apprentices were required to do this cutting up either by power saw or by shearing machine. The blacksmith then took the cut lengths and first of all rounded the end, and then punched the hole. To keep the strength round the hole, a slot was punched and then opened out, and a mandrel the exact size was hammered through the hole. When all the holes were punched it was usual to have a tea break or a smoke, and then they were turned round. The next operation was to swage the end to a smaller diameter and then turn the end at right angles.

The final bend was very important because when closed up cold, the round end had to fit the hole exactly or else it was of no use. This was the least time-consuming operation and was always carried out by Father. Eventually I was entrusted with this operation.

When the links were cold it was the job of the apprentice to sort them out into their different sizes, wire them into dozens and then pack the orders into bags ready for dispatch by rail. The railway wagon collected goods from the mills in Rastrick every afternoon, so all we had to do was hang a notice on the railing in Ogden Lane - LMS to call – and they were on their way.

The job was really our bread and butter, as we used to say. It was not to last however. One day some boffin in London concerned with health and safety had a rush of blood to his brain and decided that any broken chain in future should be repaired by an 'atomic welded' link, the same shape as the original chain, and that the chain should then be tested to twice the load that it could legitimately carry, and a test certificate be issued.

The Riley link was killed off at the stroke of a pen. It was illegal. Of all the thousands that we had made, we never had even one complaint that the Riley link had broken.

A Village Blacksmith

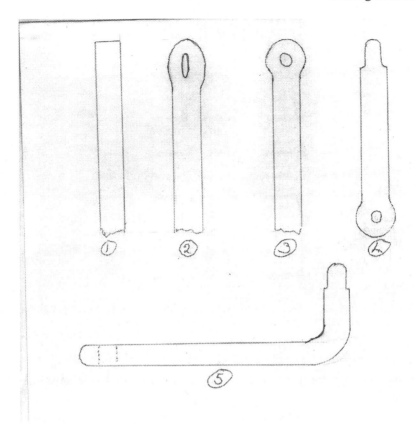

<u>Riley Link different operations:</u>
1. Cut length from round bar 12ft long.
2. Round in the end with the hammer and then punch the slot (we used to shape an old file for this as it was very hard material).
3. Using a round punch pull the slot into shape, then knock a round mandrel exactly the right size through the hole.
4. Turn the piece round and reduce the end in size to fit tight into the hole just punched.
5. Bend up the end at right angles to the exact length (we had a gauge for this in the anvil). After reducing the end it had to be quickly cooled off in water to stop it being damaged with hammering. It was now ready for the final bend.

Chapter 25

ICI work

One day we received a telephone call from the Engineering department of the ICI Huddersfield. Would I go and see Mr Evans, head of that department? I felt a bit apprehensive as I announced myself to the gate man, but he was very helpful. As I did not know my way about the works, a telephone call brought a messenger to the gate, and he escorted me to the office. It was explained to me that they had just closed down their own blacksmiths shop and were looking for a firm to do a good all-round job with prompt service when urgency was required – i.e. if a plant was stopped and a part or repair was most urgently required. I said we would do our best to meet their standards.

He gave me a standing order for some forged eye bolts, which were very easy for us, and they were impressed with our prompt delivery. I was then shown a drawing of a much larger job to test us out. The items were called slicer bars. The shafts were 1 ¼" in diameter and 12 feet long, bent at one end to form a handle. On the other end there was a forged piece like half an arrow head. These were for ramming into the boilers and removing the clinkers. This end was 12" long, 5" deep at one end and 1 ¼" thick. The edge was forged to make a cutting edge.

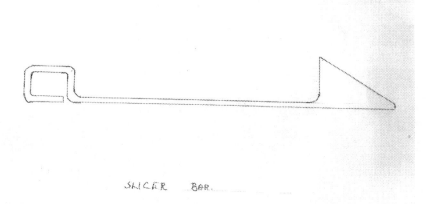

Slicer bar

The pointed end was welded to the shaft and this was to test our welding as well as our forging. Again I got the ok and was very pleased. That was the start of a long association with ICI. A memo was sent to the builders' yard, electrical department and tinners' shop, that G.S. Whiteley were now official suppliers and would see to their requirements in the blacksmith line. Soon I was called to the builders' yard and came away with a load of pneumatic tools to sharpen. This carried on for years. Next I was called to the electrical department. Their jobs were easy – it was all brackets for hanging cable and they gave me an official order for the next 12 months.

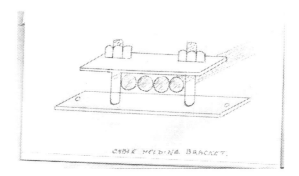

Cable holding Bracket

One set of brackets comprised a plate with 2 protruding threaded screws, and a plate with 2 holes to fit over the top for holding cables.

We had been doing these for years by the same method. We drilled two holes in the plate and inserted a countersunk headed screw and welded it over at the back. One day I saw at an exhibition a new stud welding gun. It was connected to the welding plant, a threaded stud was placed in the gun and the trigger pulled. There was a flash and the intense heat welded the stud to the plate. The studs were cheaper to buy than the countersunk screws, and it would save drilling and CSK[23] 2 holes.

---

[23] countersinking

When you are supplying these in batches of 100, there is quite a saving to be had. I immediately ordered the gun and 1000 studs to weld on. We made a jig for setting up and awaited the next order. It never came. I had to visit the builders' yard with some tools, so I thought I would call and see the store man in the electrical department to see if he required any more plates. The building was closed and was being demolished. They had decided to put all electrical work out to contract. The machine was never used and all 1000 studs were scrap.

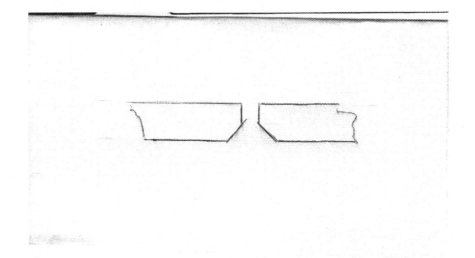

Countersinking. This means to open out a hole so that a screw fits flush

That was ICI. The unions were very strong at ICI and whenever I delivered any goods to the engineering department various men always examined them. I thought it was just curiosity but it was far from that. One day I received an order for some very small forgings – they were only 4" long made from 1 ¼" diameter. They had to be flattened at one end. They were so small we kept losing them in the fire and there was not enough length to grip properly to flatten them under the hammer. I had a brainwave. I decided to cut lengths off and then mill the sides off on a milling machine. This I did and they were a lot better job.

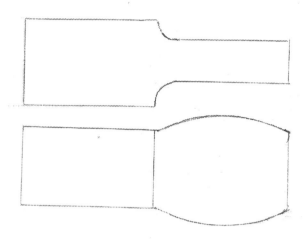

Too small to hold, but big enough to cause a strike

The day after I delivered them I received an urgent phone call to go to the mechanics shop as soon as possible, to avert a strike. They saw that they had not been forged but machined. I was a blacksmith, and this was stealing work that they could do themselves. I had to apologise and bring them back. I put them in the furnace, got them red hot and let them scale over and then took them in again.

The militancy of the trade unions worked to our advantage in another part of the works. The men on the 'plant' (working with the chemicals) often needed to open up certain vessels. These had a lid which was fastened on by several eye bolts fastened to the end of the tank. These were swung over and into slots in the rim of the lid, and fastened tight down with hexagon nuts. The trouble was a 'plant' man was not allowed to use a spanner as this was an engineer's job. They had to send down to the mechanics' shop, for a man to go with a spanner to tighten or slacken the nuts as the case may be. This was very time-consuming and could often hold up production. One day one of the bosses saw what was happening and came up with a great idea. He took off all the hexagon nuts and replaced them with 'ring nuts'. These could be tightened by putting a bar of metal through the ring, and using this as a lever. The job was complete without using a spanner. The engineers could not complain if the ring nut had been tightened by a piece of bar that just happened to be lying about! Soon all the vats and tanks were to convert to this method. This was a job for us and we supplied hundreds of them. News travelled fast and soon we were asked to supply these ring nuts to the ICI at Wilton in Cleveland.

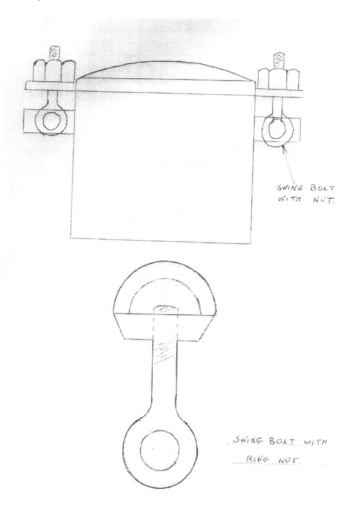

Swing Bolts and Nuts

The ICI was a good firm to work for insofar as money was concerned. All jobs had to be quoted for (except emergency repairs when speed was of the utmost importance) and an official order supplied. It could be for 1 delivery or could cover goods for 1 year. If the invoice was submitted at the right time, and all the relevant numbers were correct, the cheque arrived on the 1st of the following month. On the other hand they could close a department down without notice and you would find that the orders just stopped. This happened with the tinners' shop as well as the electrical department. They used to make pumps from sheet steel for taking samples out of barrels or tanks, and we made the plungers with the handle end forged. One day when visiting the works with some other goods, I noticed the tinners' shop was gone. One would have expected the ICI to go on forever but often conditions change, and overseas firms work cheaper. Whatever the reason the ICI was taken over, and changed its name about 3 times up to the time of writing.

Chapter 26

The Flood

It was in the late 1960s when we had THE hailstorm. Looking back it was nothing compared with some of the events that have happened in the years 2004-5, but at the time it was a big event. It was at the end of summer one day and the sun disappeared behind a cloud and gradually the sky darkened and almost turned day into night. Then there was the first crack of thunder and the sky lit with flashes of lightening. It started to rain, then it poured and then there was a drumming sound as hailstones came hurtling down. We had to run for cover although all the doors in the works were wide open at this stage – we saw no reason to close them. We stood watching through the open doors and by now water was gushing down Toothill Bank, Ogden Lane and Bowling Alley. The hail persisted and hailstones the size of golf balls swirled about in the water, which had turned into a torrent. Bottom of town was flooded and water cascaded over our boundary wall, down the yard and was trapped at the bottom of the yard by the golf course wall. The flood water reached over 3 ft high before the storm abated as quickly as it had begun.

We were helpless as we stood and watched the water rise. The van was filled with water to the level of the seat. It was ruined. The power saw which we kept outside was covered and the motor wrecked. As the water seeped away we were left with a carpet of hailstones 3 ft deep. They were level with the benches. The whole yard was carpeted with them – we thought they would soon melt, but melt they wouldn't. The generator for the D. C. welding plant was ruined; all ours and our customers' drawings were sodden. All tools in drawers and cupboards were ruined. Micrometers had rusted and were useless and everything was slimy with all the mud and debris that had been washed down. All the stock of electric welding rods was ruined. The little cottage at the bottom of the yard had the old coal shoot still in place with its ill-fitting door (which just hung under its own weight). The cellar was flooded and solid to the ceiling in hailstones. The machine shop was similarly affected, being about 3 ft deep all through.

We had round 'Romesse' pot-bellied stoves in both the welding shop and the machine shop, and we stoked these up until they were glowing red. All that happened was, they melted the hailstones for an area of about 12" round the stove and that was it. The stoves had no effect whatsoever on the rest of the buildings. There was only one thing for it, we had to shovel the hailstones into barrows and wheel them away. That was after we had cleared a way down the yard. With everyone working hard it took us two weeks to get cleaned up before we could do any work. Weeks after, people could not believe what had happened and thought hailstones like golf balls was a fantasy, until we went into the cellar and brought out the evidence. It was Christmas before these disappeared.

We contacted the insurance man to see what he could do for us.

> "Nothing", was the short answer. "If I had suggested that you insure against flood, a mile above the river, you would have chased me out of the shop."

He was probably right. The event cost us a lot of money.

Clifford Riley

On my way home that day I had to walk, and at the bottom of Clifton Common the beck had overflowed and I had to wade through 2 ft 6 in of floodwater. There were some children waiting for help so I carried them to dry land in their turn. When I got home I did not have a flower left in the garden, and the shrubs were stripped of their leaves. Park Church, where I was Trustees' Secretary, had 50 windows broken.

Chapter 27

The French Connection

It was the summer of 1971 when my nephew John arrived at the works on his bicycle. He came straight to the point.

"I've come to see if you can find me a job during the school holidays, Uncle Clifford", he said.

I was delighted, but I said,

"Well, I suppose we could find you something to do".

I suggested what we might pay if he worked full time, and his face lit up and he said,

"Great, when can I start?"

"Now, if you want".

That shook him a bit, but he propped his bike against the wall out of the way and said,

"Right, I'm ready".

One reason I was so pleased to see him was that I had recently quoted one of our regular customers for a job, and the order came back by return of post. The job consisted of a lot of girders and stantions, plates and trunions and tie bolts. There were hundreds of holes to drill and all items had to have a coat of paint. The drawings we had to work with (and quote from) were for a French firm in the end, and all measurements and comments on the drawing were in French.

I had not been happy with my quotation because I had great difficulty in picturing how the whole job fitted together. However, we made a start – Joe, myself and John. It was 'hold this, John' and 'fetch that, John', 'drill this, John' and 'take this and do that with it'. We had another pair of hands which were so useful. To his credit, John never stopped, never complained and just got on with the task. We tired him out but he genuinely enjoyed it and was there in good time the next day. It soon became apparent as we finished various parts, that something was not just right, so together me and Joe set out the drawings and discussed them in great detail. Then the answer came – I had missed something on the original drawing and had only quoted for half of what was required.

Catastrophe!

I had to order more steel and carry on in the knowledge that at the end we were going to lose a lot of money. It so happened that once we had the 'hang of it', as we say, the job flowed through the shop like a dream, there were no snags and everything fitted perfectly. Our customer was very generous in his praise for a job well done. It's very grievous when you do a job like this and lose pounds!

A Village Blacksmith

Without John our loss would have been a lot worse. He could not have arranged his baptism at work at a better time if he had tried.

John at work

Chapter 28

John

John at work

John started work full time in 1973. He had shown how adaptable he was when he was working during his holidays from school, but we never talked about long term employment. To be honest I had no idea he would want to be a blacksmith, but I was really pleased when he came to see me and said he had given the situation some thought and decided he would like to work with his hands. It transpired that, apart from the practicality of the work suiting him, John was attracted to the sense of traditionalism that permeated the firm. The fact that the business had passed seamlessly, from uncle to nephew, father to son struck a chord with him and he enjoyed being a part of that continuity. With the experience gained during holiday working, he just fitted into the routine right from the start.

He went to Halifax Technical College to learn welding and passed his City and Guilds without ever welding anything! This was no fault of John's, but I was very disappointed with the whole set up and vowed never again to send an apprentice to Tech, to be paid for a day off as well. We had a very good reputation for welding and it was not easy for John to learn welding in the shop. He had to know how to weld so I arranged for him to attend the British Oxygen School of Welding in Leeds, where I myself had been many years ago. Here he had expert tuition in both electric and oxy acetylene welding, and was then able to tackle proper welding jobs with confidence.

It was some time later that we had another encounter with the factory inspector. All had gone well until he noticed the fork-lift truck on charge. Unfortunately the door had been left open, so it was there for him to see.

"Who is the authorised fork-lift truck driver?" he demanded.

"Whoever needs it uses it as we normally just bring in plates to put on the cutting machine", I said.

"That is against the law", he spluttered, "you must have one man designated to be official driver and he must have a Ministry of Transport certificate of competence. You must not use that fork-lift until that rule is complied with".

He gave us the name and address of a place in Leeds where the designated person should attend. Of course I designated John. This course took about 2 weeks. This seemed mad when everyone in the shop could use the fork-lift for all that was needed. It had 1 lever for forward and reverse, 1 to go up and down and a foot pedal for go. How could anyone need 2 weeks' tuition? Poor John. We dropped him in for it. He had to train on all sorts of different machines, learn about stacking and retrieving, how to load a 20 ton wagon or trailer, and a bit of maintenance. It was a very exacting course and John was very worried that he would not pass. But of course he did, and proudly came back with his certificate.

People used to ask the old question, 'do you shoe horses?' The answer was no, and we never did. In my time there were two farriers in Brighouse so there was no need for another, and our main trade was quarry tools. Soon after John got married to Wendy he got interested in horse riding. So much so that one day he suggested that we might shoe horses. Horse riding was becoming very popular, and there were now no farriers in the area. It came as something of a shock, but I didn't want to dampen his ardour. I had visions of horses out of control and piles of you know what about the place. When I asked just where he would do the shoeing, he said if I removed my caravan from its shed that would be just the place, and there was room to tether them alongside. He had obviously been mulling this over in his mind for some time, and had all the answers. As for myself, I was not convinced, saw all sorts of pitfalls, but promised to think about it.

I had not come to any specific conclusions when I got a telephone call to the effect that John would not be in on Monday – he had hurt his back. It transpired that over the weekend he had been riding and had come into contact with a low branch. He was swept off the back of the horse and was lucky that no limbs were broken. Shoeing horses was never mentioned again.

John was now taking more responsibility and dealing with customers as they came for their orders. He had a very likeable manner and customers took to him. In time they would come and ask 'is John about?' In 1980 John was made a Director of the firm and with his share of the profits purchased shares in the business, so that on my retirement he owned the majority of the shares and the business just carried on as usual.

Clifford Riley

Clifford and John

My father came in and out of the shop until he was in his 80s and often we had to do his jobs over again as his eyesight was failing, but work was his life and he only lived for 2 weeks when he actually gave up. I promised John this would not happen to him. I told him to do his own thing, and I would not interfere in any way. When I walked out of the works at Easter in 1986 I did not go near the works for 3 months to give John time to settle.

Two Pictures of the Smithy, about 70 years apart

c.1920 – the blurred figure is John Riley

c.1990 – John Snell far right

Chapter 29

Nora's Death

It was a terrible blow when Nora died suddenly in December 1982. We set out together on the Monday morning to attend my cousin's funeral. During the service in the cemetery chapel, she suddenly put her hand to her head and said she felt unwell. She bent over and put her head on the pew in front, and managed to get to the end of the service. People further along climbed out over the pews and proceeded to the graveside. I fetched the car and, with the help of another cousin, got her into the car, and home. We were living on New Street at Clifton at the time. I phoned for the doctor, and managed to get her into bed. The doctor arrived and seemed unconcerned at her condition – 'a bit faint with the funeral', he observed.

"If she is not alright in the morning, give me a ring".

She lapsed in and out of consciousness and an hour later she died in my arms from a brain haemorrhage. You have to have this experience to know what it is like. No-one can tell you.

A lot of prayers were said for me, I know, and when the situation seemed hopeless I received a telephone call from Joan, a lady who worked for one of our customers and was quite familiar with our work. She told me she had been made redundant and needed a part-time job. She was sure I must need some help. I sure did, and she started the following day and ran the office like a smooth-running Rolls Royce. I couldn't believe my good fortune. Another call followed from a lady I knew who offered me two half-days a week to keep my house in order. She had previously worked for my aunt, so I knew she was ok. She looked after me until I married Emmie the following year. Joan was still in charge of the office when I retired.

Mick, Eric, Joan, Clifford and David

Chapter 30

Uncle George

I'm not sure how old our premises are, but like many old buildings we had our ghost. We called it Uncle George, after the founder of the business, who was George Shaw Whiteley. You may or may not believe in such things, but I assure you that things have happened which defy all other explanation: the hammering when no one is there, the cold room, the disappearing objects and the noisy disturbances.

I know very little about my father's Uncle George, except that he founded the business in about 1860 and that he played the violin. His full name was George Shaw Whiteley, and the firm continued through the different generations trading as G.S. Whiteley & Co. My father eventually borrowed money from an aunt, and bought the company in 1910. After he died I formed a limited company and we became, G.S. Whiteley (Brighouse) Ltd.

Before I describe some of 'Uncle George's' antics I will explain how the forge was set out. The works were a set of separate buildings, known as 'top shop', 'middle shop', 'long shop' and 'welding shop'. There were 2 forges in the top shop, and three in the middle shop. The long shop housed the engine, lathe, drills, saw, grinding machine, power press, and 1 forge and hammer. Also a bar shearing machine and metal racks. As the works were on a slope, the middle shop was entered from the top shop, by a door opening straight onto 2 steps. Very dangerous really! The door had a little window about 9" square and was closed by a weight on a chain. Most of the time the glass was shattered by the door clanging, and was replaced by a piece of wood or sheet metal.

My father, who was always referred to as the 'old man', worked at the big hearth in the top shop, and had the use of a power hammer. An apprentice had the second, smaller hearth and from time to time shared the hammer. I worked in the middle shop, where we had 2 large hearths and 1 small one, and 2 power hammers.

During the War, in 1939, the premises had to be blacked out, and with the old buildings having air holes for the fumes to escape it was quite a job. One got used to working with the doors wide open, but after black out it became very claustrophobic. At that time we were working from 7.30am till 7.30pm and as the apprentice could not work overtime and the old man did his bookkeeping in the evening, the top shop was often empty from 5 o'clock till 7.30pm. One evening when I was doing hand forging with no hammer running, I heard someone apparently sharpening a chisel in the top shop.

> "That's funny", I thought, "the fire there will have died out by this time".

I glanced through the little window. There was no one in sight. I opened the door, mounted the 2 steps and was amazed to find the place empty. All was quiet.

> "I must be dreaming", thought I, and continued with my work.

Not so. Again the same measured beat on the anvil. One can interpret the different sounds and timing of the hammer, and have a good idea what job is being done. This was not the blow of hitting white-hot metal as in a forging. This was the steady ringing tone of hammer on hard metal.

I did not mention this episode until something happened again. When it did, I related the experience to my father, and asked him where Uncle George had worked in his day.

> "In the big hearth in the top shop", he replied, "that was always the boss's fire".

> "Well that explains it", I said, "he's still around and sharpening his chisels".

> "That's strange", said father, "that was what he always did".

Thus was born the legend of Uncle George, the friendly ghost who from time to time came back to keep in practice.

I have never seen a ghost, but yet I believe that something happens from time to time which defies explanation. Whatever happened at work was thereafter attributed to Uncle George. The hammering continued for some time, but was never heard again after the war when the blackouts were removed. That was not the end of the 'happenings', however.

There was a little cottage above the top shop, which was taken over and used as a store place and later on as an office whilst the new office was being altered. This joined onto the next cottage up Ogden Lane. It was here where a young couple had lived for about 6 months. One morning, as I was on my way to 'open up', the young man was waiting for me.

> "Mr Riley, I fear you have had a break in. There were noises coming from here in the night of things being dragged about. We couldn't sleep. I'll come with you to see what damage has been done if you wish".

I accepted his offer and in we went, up the stairs and into what was formally the bedroom. Everything was as normal. Not a thing out of place. The young man couldn't believe his eyes.

> "How can you account for that?" he said.

I pondered a while, and then explained about Uncle George, our friendly ghost, and told him not to worry. He did not look convinced. I told Nora about this when I got home.

> "That was a bit silly, wasn't it?" she said. "He has two options now – he will either believe you or flit".

He flitted.

At the time, my niece Ruth and her boyfriend had finished at University, and were looking for a little place of their own. I jokingly mentioned that the cottage in Ogden Lane was empty, never thinking for one moment that they would be interested, but of course they were. I did mention about Uncle George but assured her he was quite harmless. They stayed for about a year, but not only did they hear the noises next door, they felt a presence in the bedroom and on many occasions they snuggled into bed and pulled the bed clothes over their heads.

So far the happenings were all in the old property round the top shop, but the next one occurred in the welding shop. We had received an order from the Ministry of Supply for some hot pressings, which had to be very accurate and be exactly alike. Joe spent the day making the press tool, which was only about 5" across and was easily held in the palm of the hand. He completed the tool late Friday afternoon, just before we finished work. I looked at the finished article along with Joe, decided it was ok and that we would proceed with the pressings on Monday morning, when I would arrive early and light the gas furnace.

We left the press tool on the welding table. On Monday morning it had vanished without trace. The premises were intact, there had been no break-in, but the tool was missing. A thorough search failed to find it, and we had to make another one. Several months later it turned up on the bench at the opposite end of the welding shop and was showing signs of rusting. Had Uncle George moved out of the old property?

Things went quiet after this for a long time, but Uncle George had not finished yet. This time he was back in the old property, in the middle shop. It was usual at the end of the day to put the ear protectors (ear-muffs as we called them) over the anvil pike where they were ready for the next morning. They had a black band over the head, and yellow muffs covering the ear, making them quite conspicuous. One morning John seemed a bit agitated as he accused Mick of pinching his ear-muffs. Mick protested his innocence and produced his own as proof. John then enquired if I had moved them or borrowed them, and of course the answer was no – mine were there on the pike as I had left them.

There was no doubt that John's ear-muffs had disappeared, and I got him another pair from the stores. About a month later, exactly the same thing happened to John again, so there was no option but to go and get another pair for him.

So seriously was Uncle George taken that, a few months before John and Wendy married, he and two of his friends decided to lie in wait for him, to see if they could catch the ghost performing his mischief. The three of them, fortified with alcohol, took their sleeping bags into the store-house one cold, Spring night. It was pitch black and the old place creaked and moaned as it settled, but, unfortunately Uncle George declined to join them. All they got for their trouble were stiff necks and aching backs from trying to sleep on a stone floor!

For the next few months all was quiet on the supernatural front, and then it happened again just as before. John was getting exasperated; he stood in the middle of the floor with his arms outstretched and, looking upwards (as if George was hovering somewhere), he said at the top of his voice,

> "Uncle George, you have had your fun, now will you please, please give us a bit of peace".

We never heard from Uncle George again, and there were no more happenings.

Chapter 31

The 'Bending' machine

We had two bending machines, where you put a straight bar of steel in the jaws, and then a moving jaw with a big long handle attached could be pulled round and the bar bent to any desired angle. I suggested to my father that we really needed a larger one for some of the work that we were now doing. He said that he would think about it.

It so happened that at that time, I received a telephone call from a firm who made 'bending' machines and had one in particular which was new on the market, and would be just the thing for a small firm like ourselves. I asked about its capacity, wondering what size of bar it would take.

> "We have a rep in the area, and if it is agreeable he will call upon you during the next few days, to demonstrate".

> "Good idea, look forwards to seeing you".

I told Father of the arrangement, and for once he was in complete agreement.

Soon after a rep introduced himself, saying,

> "I have the new machine to demonstrate".

"Good", I said, clearing a space on the welding table, "we'll put it on here and see what it will do".

Father came to look too as the van arrived and the rep appeared struggling with a big cardboard box. I thought, "This doesn't look heavy enough for what we want".

But he put it down carefully and proceeded to get another smaller box from the van. He came in all smiles and proceeded to open the box to reveal a VENDING machine!!! The smaller box contained packets of tea, coffee, cocoa, soup, milk and cups etc. We looked in astonishment and then realised the mistake.

John, Mick and Joe, sensing that something was happening, all gathered round while the salesman went through his spiel. Of course, they did not know of my mistake and, thinking the 'old man' was going soft in his old age, showed great enthusiasm and congratulated Father for providing something that we could have done with for a long time. He was taken aback, but did soften, and thus G.S. Whiteley came into the modern age and tea breaks became official.

Everyone had a good laugh about the whole situation and thought it a great joke. Later, we went to an engineering exhibition, and did get our Bending machine as well.

Clifford Riley

Mick, Eric, John, Clifford and David

Chapter 32

Emmie

Emmie was a member of the church I attended so we knew each other in a casual sort of way and always exchanged a word when our paths crossed. One day my sister in law Marj and her husband Eric were having a chat with Emmie and it transpired that she often visited two old aunts in Morecambe. This was a coincidence because I used to take Marj and Eric to see a relative who lived near the Battery Hotel. Emmie travelled by train.

"You must come with us next time", said Marj as she volunteered my services!

A date was arranged and we set off together. We had a meal at an hotel and then visited our separate relatives. As Emmie's relatives lived at Bare, I dropped off Marj and Eric first at one end of Morecambe and then proceeded to take Emmie down to Bare. This was the start of a more intimate relationship and we were eventually married in October 1983.

Emmie

Chapter 33

Praise Indeed

It is always a pleasure to see work that you have done ages ago still there for everyone to see years later. The weather vane, gates and railings, simple door grills, church flower stands and candle holders all spring to mind. Sometimes a surprise hits you, out of the blue as we say.

I was on a holiday in the Lake District, taking the usual Easter break, and it was raining and overcast. The friend that I was staying with suggested we do a car ride away from all the usual haunts like Windermere and Keswick and take to the bye roads of which she was well acquainted. Out in the wilds over Honister Pass we came upon a slate mine; nothing unusual about that, except that there was a sign outside which said, 'Open to the public today'.

"Are you interested?" asked my friend.

The answer was yes and in we went.

The actual mine was closed, but we could walk round the warehouse, see all the machinery and purchase all kinds of souvenirs made of slate. The person in charge was demonstrating the art of splitting slate. Using previously cut, 3" thick, circles and square blocks of slate, he showed us how to make thin teapot stands and engrave patterns on them. At the end of the demonstration he gathered a few people round him and gave them a chance to take the hammer and chisel and have a go.

I worked my way forward and waited till last before picking up the chisel. There, plain to be seen, on the shaft was my stamp, RILEY, which was to be found stamped on all our chisels. I introduced myself to the man giving the demonstration, who turned out to be the proprietor. He could not believe the co-incidence. We had a long chat and he was full of praise for our chisels. I asked him where he got them sharpened in this out of the way place.

He laughed.

"Never had them sharpened, they go on for ever", he said.

Such praise is praise indeed.

It was some time later when I was watching television that, by chance, the Chelsea Flower Show followed the programme I had been watching. I became interested and continued to watch. One particular exhibit, which was drawing a lot of attention, had made a lot of use of slate in the arrangement. In a flash back the film showed the lady in charge of the exhibit discussing her requirements at the slate mine in the Lakes. There he was, the proprietor, splitting and cutting her slate to specific instructions, using my chisel. The exhibit won a prize.

"That was my chisel he was using!" I shouted to the television.

Chapter 34

Retirement

Clifford Retires

My Brother-in-law Max could tell you how many months, hours and minutes to the time when he could retire and leave work behind him. I never had that feeling; in fact mine was just the opposite. I dreaded the day when I would hang my leather apron up for the last time. My father pottered about the works till he was in his 80s, but I had promised John that when he took over I would not interfere and I kept my word; never going near the works again for about three months after my retirement on March 16th 1986.

What was I to do? The lawns were cut, the hedges trimmed, the roses pruned, the car and caravan cleaned and I had only been retired for three weeks! When I first married I inherited an old greenhouse and had the pleasure of enjoying home-grown tomatoes. 'That's it', I thought and purchased a greenhouse, proceeding to erect it and get it planted out with tomatoes.

I remembered that a former customer had a vine in his greenhouse and could always produce a bunch of lovely grapes. Off to the nursery we went and came back with a black grape plant. It filled the greenhouse from ground to roof, cut out the light, but never produced a single grape!

> "Why not grow your own vegetables?" said a neighbour, "They always taste better straight out of the garden."

We had a paddock that was about a third of an acre in size, so I marked off a plot near the house and proceeded to dig a vegetable plot. That gave me back ache, so was not really a good idea. Strangely the soil and the weather must have been just right, as soon green shoots were springing up in nice long green lines. Mercifully after about 6 weeks the rabbits came and cleared the lot, saving me from further backache.

As summer turned to autumn we made good use of the caravan. I was now beginning to relax a little. Sadly, that year, Emmie, my second wife, was diagnosed as having cancer and spent Christmas in hospital. That put a different aspect on retirement.

I had joined Clifton Handbell Ringers in 1977 after the old Clifton bells had been found, but they were in a terrible condition. Along with a few other members I assisted in repairing about 50 broken leather handles and also welded cracks in a couple of bells. This gave me an interest, not only in the ringing, but also in the repair and maintenance of the bells.

It was about this time that the vicar of Honley Parish Church, near Huddersfield, decided that their old hand bells were of no further use and he put them up for sale. A local dealer made an offer, which was accepted. Unfortunately they were in a terrible state and no-one would touch them, so the dealer was 'landed with them', as we say. After he had given up all hope of selling them I heard from a fellow bell ringer that if he was made an offer the dealer would be glad to rid himself of these 'blessed' bells.

> "Why don't you make an offer and then do them up?" said my friend, Peter.

This I did and became the proud owner of 86 hand bells.

I had had a little workshop in the attic for some time, for doing odd jobs, but now I could put it to good use. I spent five years of my spare time refurbishing these bells and then made boxes for them to be stored in, all lined with foam. When I had finished they were valued at £17,000. I derived much pleasure from doing the work and also hearing them being used, both by the church and for Christmas ringing with Clifton Handbell Ringers. Eventually when I moved into my flat there was no room for them, so I presented them to the Lord Mayor, Trelor Home, for disabled children at Alton in Hampshire (a charity I had supported for over 50 years).

Clifford Riley

When Emmie died in 1997 I was left alone in a big bungalow with a big garden to match, so when I had the opportunity to buy one of the new flats in Knightsbridge Court I had no hesitation in leaving Clifton for a new adventure. It took me a long time to realize that I did not have to be doing something all the time and that there was a time for relaxing, which wasn't a waste of time. Now, after a heart attack and the onset of arthritis, I can fritter time away with a clear conscience.

Relaxing at home

Chapter 35

Smithy's End.

By Wendy Snell and Clifford Riley

Wendy Snell, c.1986

After Uncle Clifford retired, John wanted to keep the family feeling of the firm, but at the same time make some of the tools appeal to a wider group of people. He had no plans for doing things differently, but changes had to be made that were desirable at the time.

Strangely the first additions came in the office and not the works. The old typewriter was replaced with an electric one and a fax machine was installed; the latter proved very useful as drawings and sketches could be exchanged between the customer and us, queries could be settled quickly and time was saved.

We had made large tungsten tipped chisels (3"wide) for a long time and had supplied forgings to a tool maker in Brighouse who specialised in smaller chisels. For this reason we had not sought to expand our trade in this direction and be in conflict with our customer. Our larger chisels, 3" nickers and pitching tools were much admired in the trade and we were constantly asked to supply the smaller type of chisels. This we decided to do.

We soon realised that our small milling machine was too small to cope, so John purchased a larger one. With this machine a jig could be set up and many tools milled (slotted for the tip) at one setting; thus saving much time. The small machine was still useful because we could use this for odd jobs without dismantling the set up.

The tips were brazed into the slots with silver solder. It was imperative that the metal was clean for this process and oil left on the tools from the hardening process had to be removed. Customers were getting choosier and oily shanks were not acceptable. One solution was to sandblast the tools, as this would not only clean them but would give them an attractive finish which, if required, could be painted for selling in shops.

The question was where to put the sanding machine. John's idea was to use the garage at the bottom of the yard where Clifford parked his caravan. When John put the idea to Clifford he pointed out how convenient it would be to have the caravan in the drive at home near to the house where he could keep an eye on it. After laughing at John's cheek he towed the caravan away!

As often happens, once word gets round that you have a specialist machine you find other uses for it and customers often brought in things that needed sand blasting. John started to put the tungsten tips on the smaller chisels and nickers and began painting them shiny black. They were more expensive than the original, but lasted much longer and looked more up to date.

The board of tools used for marketing purposes

All Whiteley's chisel heads were turned in the lathe, but the old lathe was very limited in its capacity. When we made scrappling mauls from 2 ½" square bar we had to make a square face at one end and a rounded face at the other end. This we did by grinding. It was dirty work, expensive on the grinding wheels and the men had to wear masks. Worse still the factory inspector needed the wheels guarding in such a way that it was impossible to grind the faces without removing the guard. There was only one thing to do and that was to look out for a big lathe. After contacting one or two machinery dealers, one was found that was just suitable and another purchase was made.

I think the best way to describe the business was 'steady'. There were never huge orders from people, but the same companies would turn up, month after month, knowing they would get high quality service, as well as good tools.

John also continued with wrought ironwork. He enjoyed this line of work as it paid well and left a permanent reminder of his capabilities. There is a lovely gate and fencing at the top of Toothill Lane in Rastrick that John made and it stands to this day.

Wrought iron pot holder made by John

Whiteley's was justly renowned for making tools for the building trade, Chisels, nickers and mallets were all good sellers, as were tools for scraping out old pointing from brickwork and smoothing the new back in again. But there was still the odd householder coming in wanting their garden shears sharpening. That was an aspect John liked about the place. People felt they could wander in and ask for help. Whiteley's wasn't a huge faceless consortium; it was still family-run and friendly.

In 1992, John began to get frequent visits from the Environment Department of Calderdale Council. These were initiated by a new neighbour who complained of the amount of noise generated by the smithy and the fact that her ornaments bounced on their shelves when hammering took place, and the noise wakened her baby. We begged the question, why move so close to a blacksmiths? Whiteley's had been trading for almost one hundred and fifty years, but she boasted to the neighbours that she would get us shut down.

The environmental officer arrived with his decibel meter and upheld her complaint, even though we had been there since 1860. He said the law was on her side. John was instructed to silence the hammers as far as possible. He ordered us to put rubber cushions under the hammers. This involved having to dig a large pit under each hammer, line it with rubber and concrete in the hammers both to deaden noise and reduce the vibration. It worked to a certain extent but, of course, not completely. The neighbour who had stated her desire to 'shut the place down', got her wish, though I hope she didn't wish for the way it came about.

Tragically, on September 10th 1995, John died, at the age of 38, of coronary heart disease. None of us had any idea he was ill. There were no chest pains, no high cholesterol or raised blood pressure. He rode a mountain bike frequently and wasn't overweight but all this counted for nothing. I was told the atheroma had been forming for at least five to seven years, interestingly about as long as he had been running Whiteley's.

Despite all our pain and anguish the business still needed sorting. At first I wondered if Uncle Clifford might want to come back, or if Robert, one of the employees, may want to buy the business. It was, after all, a going concern.

However, Clifford was happy in his retirement and Robert could not afford to buy, so another solution had to be found. A business consultant was hired to see if he could find a buyer quickly but to no avail. After all, how many blacksmiths are there and how many people would want to take over the place once they knew they had neighbours who wanted them out?

It wasn't an easy decision but it had to be done; the works were closed down, machinery and stock were sold and letters sent to all interested parties. Everyone (apart from one buyer) paid what they owed by Christmas of 1995, which was a big help. I also took the decision to have the buildings torn down. This was for safety as much as anything. The roofs of the long shop had always been fragile and I dreaded children playing there and getting badly injured. One thing definitely not needed was a lawsuit.

I often wonder, with hindsight (what a wonderful thing), if I should have done things differently, but I can't see how I could. No-one wanted to buy the place, it was unique and it was obvious that the Council was relieved Whiteley's had gone away. An industrial business in the middle of a residential site was no longer desirable.

It saddens me greatly that our children didn't have the chance to work with their Dad at Whiteley's. Our youngest, in particular, would have fitted right in and, maybe, have carried the firm forward into the twenty-first century. We will never know.

Chapter 36

Finale

It is impossible to condense a life story into a little book, but I hope that you, the reader, have seen and experienced something of the life of a Blacksmith in this narrative. Four generations have been involved, but I am the last of the line, so it has been rather exciting for me to look back at some of the incidents which made a lasting impression.

It was my mother's wish that I should be a musician and it is one of my regrets that I have only dabbled in music, but achieved little. I had some piano lessons when I was about 11 years old, but had a good excuse to give up when I passed for the Grammar School and started doing homework. The lessons, though, were never forgotten and gave me a musical base to work from all my life. Marrying Nora introduced me to a very musical family and it was with her encouragement that I joined her in lessons on the cello from an old family friend Harry Hampson. I was 40 years old at the time!

Like my piano playing, this venture into cello playing was not to last. We moved house from Rastrick to Clifton and before moving in we decorated throughout ourselves and made many improvements, so there was no time for music and, sadly, at the same time Harry died. Nora had been using Harry's second cello and her dad borrowed an old battered one from a friend of his for me to practise on. I did not tell my father I was playing the cello for a while, but he heard on the 'grapevine' and was not too pleased because I had not told him myself. A few days later he arrived at my door with his beloved cello and made me a present of it.

Sometime later the Sutcliffe family was invited to do a concert for the forthcoming Church Bazaar. We did several items singing and playing, but we did two pieces with us all playing instruments. In one of them I had a little cello obligato to play. I was so nervous my hand was shaking, but it produced a very moving vibrato effect. My father was in the audience and at the end he had tears in his eyes, he was so proud hearing me play his cello.

Later, as the business became more demanding, I saw no hope of playing again and foolishly sold the old cello. It had a Stradivarius label inside, but experts said it was not authentic; still a beautiful instrument though.

After a few more years the music bug began to tease me again. I bought a small keyboard and began to rehearse again. As I improved I kept running out of keys with my left hand, so I took the plunge and bought a piano. I practised on this with a lot more gusto than I did in my youth and had much enjoyment from it.

In 77, at the age of 56, I joined the Clifton Handbell Ringers and at the time of writing I am still a member. This has been a wonderful experience. We have won first prize at many musical festivals and played to huge audiences at places like Huddersfield, Durham, Wolverhampton and Manchester. On retiring from work in 1986 I took charge of a team of pensioners and we still entertain locally at Christmas. I arrange all the music for the team, which is surprising, because I was never taught how to do this. It must have been born in me.

In the year 2000 I moved into a flat and, having no garden to keep in order, I found myself with a lot of time on my hands. The music bug started biting again and I could not put away that feeling of satisfaction when playing a cello out of my mind. I decided to try again. I rang a well known dealer and asked if he had a cello suitable for a student. I did not want to spend a few thousand pounds and then find that I could not play again.

> "How old is the student?" he enquired, wondering if a half or full size cello would be required.
>
> "82!" I said.
>
> "Good God!" was the reply.

After a little conversation he said if I called at the shop he would fit me up with something suitable. I came away with a cheap one (sorry Father), made in China and, although I say it myself, I have practised hard and done very well, until I was stopped in my tracks by an attack of arthritis. I am still hoping to play again soon.

Looking back I hope that my mother, if she has been watching, is not too disappointed with my musical achievements. I did try.

When my mother died I was destined to become a Blacksmith. It has been hard work, but very rewarding, as I hope my narrative has shown. I never had a Monday morning feeling and most of the time I enjoyed my work. I took a pride in it and always tried to do a good job. I carried on the old tradition and yet moved with the times.

All good things come to an end, however and at the time of John's death circumstances over which we had no control loomed large and we had no alternative but to close the business. When my time comes it will be the end of an era. I hope people will remember John and me as a couple of fellows who did a good job.

From a newspaper article

Glossary of Tools

G. S. WHITELEY (Brighouse) LTD.

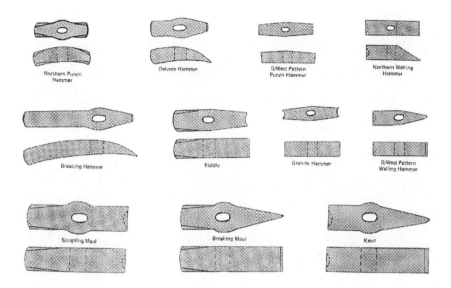

Breaking Hammer – same style as delvers hammer, but heavier

Breaking Maul – with rounded face and pointed end

Brick Nicker

Claw Tool holder

Cotter – for tightening the saw blade

Crowbars – to lever the stone

Delvers Hammer – curved with rounded face and pointed end

Drills – of different types, to make and to sharpen

Dummy Dog – for inserting in front of connecting rod on machine

Granite Hammer – special shape. Granite is much harder than stone

Holeing and Bottoming Picks – to prepare for splitting wedges

Kevil – cross between breaking maul and scrapling maul, with point and square face

Lifting Lewis – flat or half round, for lifting without damage

Lifting Wedges – for lifting from stone bed to insert rollers

Nip Dogs – sold as a pair, with chain, to lift the block of stone

Pitch Nicker – pitching tool

Planing Tools – for lintels and steps etc… (Accuracy needed here)

Plugs and Feathers – for splitting (note half holes)

Pointing Irons – for shaping the seams between building bricks

Punching Hammer – slightly curved with two rounded faces

Rock Dogs – to hammer into a seam and drag out the block

Saw Frame Dog – for holding the long saw blade

Scrapling Maul – with rounded face and square face

Sling Chains – for dragging and lifting stone

Splitting Wedges – to follow the picks

Stone Nicker – for splitting stone

Walling Hammer – used by masons, square face and pointed end

Wire Rope End – to forge and fit to new wire rope

Hand Tools included – Punch, Chisel, Nicker, Tooler, Lettering Chisels, Gauges, Lead Paring tools and Claw tools.

Printed in the United Kingdom
by Lightning Source UK Ltd.
121785UK00001B/187-276/A